The wind is a fickle source of power, with speeds frequently too low to be of sustained practical use; as a result windpower has generally remained a marginal resource. Since the inception of windpower around AD 1000 technology has been deployed to obtain the most economical power from wind. The author traces the technical evolution, concentrating on the growth in understanding of wind and charting crucial developments in windmill design. Whilst the core of the book focuses on North Western Europe, the origins of the horizontal windmill in Persia, Tibet and China are examined, as well as the widespread use of windpower for water supply in North America. Gradually, windmills were improved but were finally eclipsed by steam engines in the 19th century with growing industrialisation of the Western economies. The book concludes with an optimistic outlook for windpower, given the heightened interest in renewable sources of energy and more efficient power transmission.

Power from wind

Photo by Michael Harverson

A working Persian mill at Nish tafun in Khorasan, Iran, in 1977, with its proud owner. Further examples of horizontal windmills are discussed in Chapter 2.

Power from wind

A history of windmill technology

RICHARD L. HILLS

CAMBRIDGE
UNIVERSITY PRESS

Published by the Press Syndicate of the University of Cambridge
The Pitt Building, Trumpington Street, Cambridge CB2 1RP
40 West 20th Street, New York, NY 10011-4211, USA
10 Stamford Road, Oakleigh, Melbourne 3166, Australia

First published 1994

Printed in Great Britain at
the University Press, Cambridge

A catalogue record for this book is available from the British Library

Library of Congress cataloguing in publication data

Hills, Richard Leslie, 1936–
Power from wind : a history of windmill technology / Richard L. Hills.
p. cm.
Includes bibliographical references and index.
ISBN 0 521 41398 2
1. Windmills – History. I. Title.
TJ823.H55 1994
621.4′53′09–dc20 93-8858 CIP

ISBN 0 521 41398 2 hardback

RO

Contents

Preface

There have been many fine books written about windmills. Most of those about Britain have concentrated on surviving mills and so give excellent accounts of grinding corn and of the millwrighting necessary to build and maintain these mills and their machinery. Recently, the origins of the windmill in the Middle Ages have been studied too so that we know better how it fitted into the pattern of early mechanisation of our basic industry, food production.

I have sought to fill a gap by looking at how the windmill was developed over the centuries to obtain the most economical power it could from the wind. Therefore I have concentrated on the growth of our understanding of the wind and how sails were improved or adapted to provide the most suitable type of power output to match the equipment being driven. This has also meant looking briefly at alternative forms of power such as water and steam. However, to cover such a broad range within the confines of a single volume has meant that I have had to concentrate on the main lines of development in windpower and not branch out into its various ramifications.

I have sought to investigate the use of windpower in industries other than corn milling. This has taken me to the Netherlands, where I owe a deep debt of gratitude to Dr Jur Kingma for his enthusiastic help not only in translating from the Dutch but in finding books and windmills for me to examine. His assistance has given the work an international aspect and has enabled me to compare the industries of one country with another.

I also wish to thank many other people for their help including A.D. George, J. Kenneth Major and John Sawtell who have assisted by drawing my attention to various books and pamphlets. The library in the History of Science & Technology Department at the University of Manchester Institute of Science and Technology has been an invaluable source of many references and I wish also to thank the librarians in many other libraries up and down the country for their help.

Richard L. Hills AUGUST 1993

CHAPTER 1

The wind

The power of the wind was well recognised in antiquity, when people took advantage of it to propel ships. Old Testament writers saw the power of God in the whirlwind which took Elijah up to heaven; and in the mighty tempest which was likely to break up the ship when Jonah was trying to run away from God.[1] The nature of the wind was well described in the Book of Ecclesiastes: 'The wind goeth toward the south, and turneth about unto the north; it whirleth about continually, and the wind returneth again according to his circuits.'[2] This is reflected too in the words of Jesus: 'The wind bloweth where it listeth, and thou hearest the sound thereof, but canst not tell where it cometh, and whither it goeth.'[3] If you sit at the side of a field of young wheat or barley, you can see the wind rippling across the ears of corn, first in one place and then in another. The strength of the wind never remains the same from one moment to the next and it is not only the speed which changes constantly but the direction too. To store wind is impossible. Once it has passed, it has gone for ever. To harness power from wind has challenged people for centuries.

In spite of its changeableness, the weather follows certain patterns across the world depending upon the season of the year as well as the situation of the locality. The rotation of the earth and the temperature gradient between the Equator and the poles cause a predominately southwesterly airstream to pass over the British Isles, so that weather fronts move in from the Atlantic, bringing with them winds in the form of cyclones. The passing of strong winds in one front will be followed by the next with a period of comparative calm in between. Over the long term, these patterns remain fairly constant. Forty-two years of continuous records for Southport showed that the annual mean wind speed varied between 84 and 118 per cent of the long-term average while for thirty-seven of those years, the variation was between 90 and 110 per cent. For fifty stations in the United States, it was found that, of the annual speeds taken over a thirty-one year period, none fell more than 18 per cent below the long-term mean value and most did not fall below 12½ per cent. Elsewhere the variation was found to reach 30 per cent.[4]

In Great Britain and similar countries, the five months from May to September are rather less windy than the remaining seven and it is generally in these seven that the strongest gales occur. Through the passage of weather fronts, wind speeds can vary greatly and can rarely be predicted much in advance. Figures taken one January around 1900 on the south coast of England show these variations (Figure 1).[5] 584 hours is the total in thirty-one days so that 307 represent just over half, but some of these would have been at night and for some of them the wind might have been too strong for a windmill, so the actual running hours would have been fewer. At a wind speed of 4.77 m/sec. (10 m.p.h.) a traditional mill would be working very slowly and doing little grinding.[6]

At moderately windy places, lengthy periods of low wind speeds can occur, as was shown in a ten year period of observations in Scotland in areas normally associated with good wind regimes. At Stornoway, with a mean annual wind speed of 7 m/sec. (14.9 m.p.h.), in one August there were 110 hours or 4½ days with speeds of less than 3.58 m/sec. (8 m.p.h.); and at Aberdeen, with an annual mean of 4.02 m/sec. (9 m.p.h.), 196 hours, or 8 days, in one November.[7] These would have been periods when a traditional windmill could not have worked at all. On the other hand, there were times when the miller would have had to work day and night to make the most of the wind and he would have been constantly in his mill, taking only short naps. At the Union Mill, Cranbrook, the miller once had only three hours in bed in a windy period lasting sixty hours.[8]

There are diurnal variations, between day and night. In the summer, there is a tendency towards higher surface wind speeds during the day because the

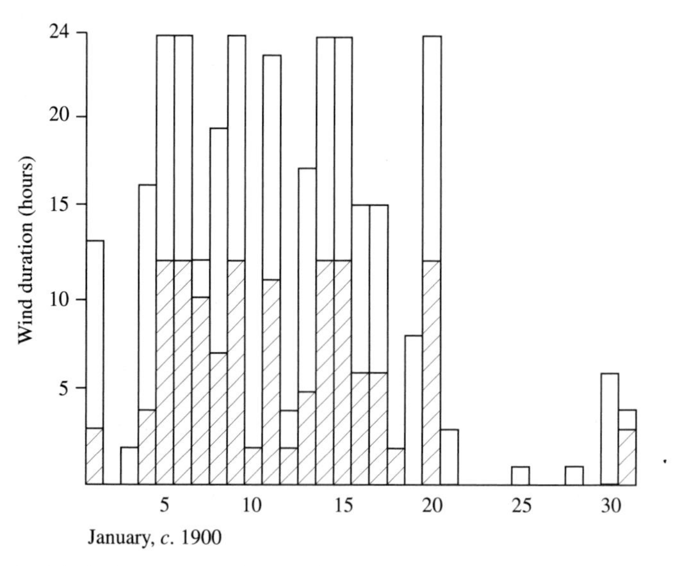

Figure 1. Graph of Wind Forces on the South Coast in January, *c.* 1900. Hours of wind □, midnight to noon; ▨, noon to midnight.

decrease in temperature with increasing altitude causes thermal convection with consequent interlocking of the air at different levels. Some of the momentum of the upper air moving at higher velocity is passed down to the lower layers, causing an increase in wind speed. At night, cooling stabilises the atmosphere, reducing convection. Then, up to 24 to 32 kilometres (15 to 20 miles) inland from the coast or even further, breezes occur due to the more rapid increase in temperature of the land during the day, causing an inflow of air from the sea to take the place of ascending warm air; while, at night, the sea does not cool as rapidly, which causes a wind from the land to the sea. These winds are particularly noticeable when the land surface is dry, for the land surface temperature is reduced by evaporation of moisture.[9]

The wind regime for a windmill will be determined both by its general location and by its particular siting. The level surface of water presents no obstruction to the wind and has little friction, so that winds passing over the sea or lakes will be the strongest and most regular. When the Dutch windmill 'Prinsenmolen' was being modernised in the years following 1935, it was considered to be exceptionally suitable for scientific investigation because it was situated close to a large lake in an area free from obstructions. With the prevailing winds passing over the lake, it was thought that measurements of wind velocity would be more than usually accurate.[10]

The areas around coasts are always the most windy, and, in Britain with the prevailing south westerlies, the strongest winds hit our western shores (Figure 2). Wind figures for Kew near London and the Scilly Isles show this difference dramatically. The figures are monthly mean velocities in miles per hour.

	Jan.	Feb.	Mar.	Apr.	May.	June	July	Aug.	Sept.	Oct.	Nov.	Dec.
Kew	8.0	8.5	8.5	7.5	7.0	7.0	7.0	6.0	6.5	7.0	8.0	
Scilly	20.6	19.5	18.4	16.1	14.1	12.9	12.4	13.9	14.6	17.2	19.3	22.0[11]

The maps of hourly mean wind speed over Britain show how the isopleths, or lines marking the same wind speeds, follow the coast. The most favoured areas are the tips of Cornwall, Pembroke, the Lleyn peninsula and Anglesey, Barrow in Furness and the west coast of Scotland. A corridor of slightly stronger winds passes from the Severn estuary to the Wash. However, the mountains, with the best waterpower sites and so less need for windmills, all lie in the west. In the lower eastern side, the need for windmills is greater because rain water soaks away through the chalk or sandstone land and the rivers do not have so many suitable sites for watermills, but it is the least windy.

Geographical features may provide more favoured sites and the contours of hills may help. A steep, smooth hill may accelerate the wind over the summit through compression of the lower layers of air by those above and so produce a wind speed 20 per cent or more higher.[12] But if the hill is too steep at the bottom, the wind may form an eddy; or if the top flattens too quickly, turbulence may occur.[13] The importance of choosing a hill with a good contour in a good location is shown below by measurements made on the top of Rhossili Down in South Wales and those at St Ann's Head forty miles away. For comparison, Leicester has been included to show an inland site.

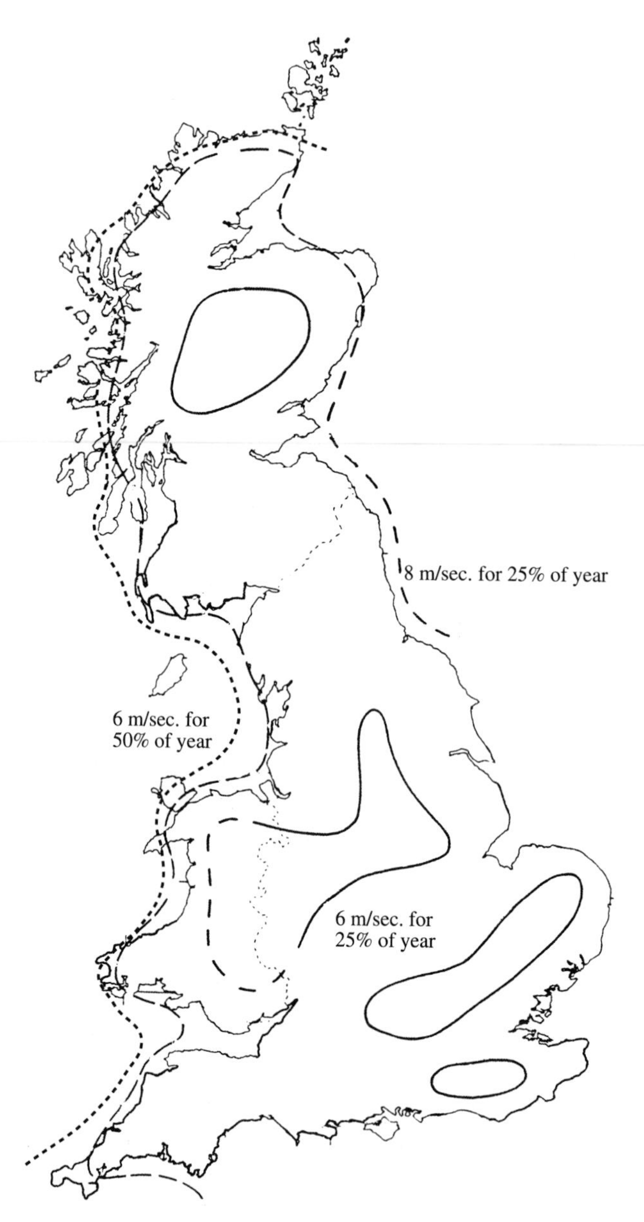

Figure 2. Diagram showing Isopleths round Great Britain, 1965–73. Areas of higher wind speed lie around the coasts and to the west. ...., wind speeds of 6 miles per second for more than 50% of the years 1965–73; ——, wind speeds of 6 m.p.sec. for more than 25%; -----, wind speeds of 8 m.p.sec. for more than 25%.

Location	Attitude, ft.	Annual average wind speed, m/sec (m.p.h.)
Rhossili	633	10.72 (24)
St Ann's Head	142	7.20 (16.2)
Leicester	267	2.75 (6.2)[14]

P.C. Putnam, in North America, found that the shape of a mountain or hill had a profound effect on the speed of the wind. A rounded hill was not necessarily the best because the wind could escape around the sides and not flow over the top. A ridge lying across the direction of the prevailing winds or nearly across them might yield the best windpower sites; but, after five years of taking measurements, Putnam and his colleagues had to admit that they could offer no reasons why the site on which he built his wind turbine, Grandpa's Knob, should have 30 per cent less wind than another similar mountain top, Seward, one and a half miles away.[15]

The immediate locality of the windmill will affect the wind. The surface of the ground causes friction with the moving air and measurements have shown that, at a level of 13 m (43 ft.) above the ground, the velocity of the wind is 10 per cent greater than at 6 m (20 ft.).[16] This has been given as one reason why the windshaft in a traditional windmill is inclined upwards a little to point into the wind, but it is based on a false understanding of the air flowing towards the ground. As long as the cloth covering the sails had to be furled or unfurled from the ground, mills could not be raised into the higher airstreams where turbulence was lower. The tall Dutch tower and industrial mills were constructed with a staging high up, from where the sails could be tended, and so their sails caught the better winds.

Vegetation, trees and houses, all cause obstructions to the wind with consequent loss of velocity. In the Cambridgeshire Fens, some of the Acts of Parliament which authorised areas or Levels to establish their own drainage systems laid down that 'no trees, stacks of hay fodder turf etc. or buildings shall be erected or built nearer to any Mill or Engine than fifty yards.'[17] The post mill at Herstmonceux, Sussex, had a conical roof over the roundhouse in two stages because the body of the mill had to be raised above an adjacent screen of trees and houses which robbed the mill of much of its wind.[18] A tree or other obstruction will cause wind turbulence, in front of it, above it and behind it. Depending upon the shape and solidity, this can be from twice to five times the height of the object in front, twice its height above and up to twenty times its height behind.[19] A shed built by the railway a few hundred yards to the south of Sandwich Mill in Kent was suspected of creating enough turbulence to be the cause of the fracture of the cast iron brakewheel. A bonfire was started to windward of the shed and the smoke was observed eddying across to the mill.[20]

The body of the mill itself will also create turbulence. In front of the mill, there will be an area of high pressure where the wind is forced around the structure. The size depends partly on the wind speed but also on the shape of the structure. Tests in a wind tunnel with a model of a Dutch windmill showed that the wind velocity could be reduced locally in this part by 60 to 70 per cent.[21] There will be a loss of power when the sails pass through this area.

A study of the power of wind assumes that the velocity can be measured. John Smeaton published a table (Figure 3) which he had received from his friend Mr Rouse, although he commented that 'the evidence for those numbers where the velocity of the wind exceeds 50 m.p.h., do not seem of equal authority with those of 50 m.p.h. and under.'[22]

Attempts to combine tests on windmills with measurements of the wind speed were made in Holland when a new type of scoopwheel for draining their polders was being evaluated. The inventor, a Dutchman, Anthoine George Eckhardt, who was a member of the Society of Haarlem and also a fellow of the

TABLE VI. *Containing the Velocity and Force of Wind, according to their common Appellations.*

Velocity of the Wind.		Perpendicular force on one foot area in pounds avoirdupois.	Common appellations of the force of winds.
Miles in one Hour.	Feet in one second.		
1	1,47	,005	Hardly perceptible.
2	2,93	,020	Just perceptible.
3	4,40	,044	
4	5,87	,097	Gentle pleasant wind.
5	7,33	,123	
10	14,67	,492	Pleasant brisk gale.
15	22,00	1,107	
20	29,34	1,968	Very brisk.
25	36,67	3,075	
30	44,01	4,429	High winds.
35	51,34	6,027	
40	58,68	7,873	Very high.
45	66,01	9,963	
50	73,35	12,300	A storm or tempest.
60	88,02	17,715	A great storm.
80	117,36	31,490	An hurricane.
100	140,70	49,200	An hurricane that tears up trees, carries buildings before it, &c.
1	2	3	

Figure 3. Smeaton's table of wind velocity and force, 1759.

Royal Society in England, had erected two mills with his new wheels which were compared with two conventional mills under the supervision of a committee of Christian Brunings, Dirk Klinkenberg and Johannes van der Wall between 1774 and 1776.[23] Brunings designed anemometers, which were built by the instrument maker J. Paauw of Leiden, with pressure plates to measure the wind force at a height of about 9 m (30 ft.). Although the instruments were treated with great care and were read systematically, neither Brunings nor his contemporaries succeeded in discovering the fundamental relationship between the readings and the power of the mills.

Admiral Sir Francis Beaufort, hydrographer to the British navy, introduced his Beaufort scale in 1805,[24] when commanding the frigate *Woolwich*. He wanted to devise a concise and repeatable method of reporting wind conditions in the ship's log for Admiralty records and based his scale on the performance of his ship in different winds: thus it was a force scale and not a speed measurement.[25] In 1846, Dr Thomas Romney Robinson of Armagh Observatory invented a form of anemometer which has become the most widely used type. This is a form of horizontal windmill with three or four hemispherical cups mounted on horizontal arms. The wind is caught in the hollow of one cup and turns the rotor while the streamlined outer surface of another cup moves against the wind. The inertia of the instrument and, in cold weather, the drag of the lubricating oil, may cause the anemometer to fail to record sudden gusts, but its simplicity and the fact that it can record winds from any angle have ensured its continuing popularity. Early wind measurements made on it were inaccurate because Dr Robinson stated that it revolved at one-third the speed of the wind, but this problem has since been corrected.[26] After the Tay Bridge, with a train on it, was blown down in a gale, the designer of the Forth Bridge, Sir Benjamin Baker (who received a knighthood for its construction) carried out a series of experiments for two years around 1880 on wind forces, using three pressure plates. The largest was 300 sq. ft. and the two others each 1.5 sq. ft.[27] As well as pressure plates, other types of anemometers available today are pressure-tube and hot-wire varieties.[28]

But even with reliable anemometers, it is still very difficult to measure the wind speeds actually hitting the rotor of a windmill. An anemometer mounted on a tower close to a windmill might not record the same winds as those reaching the mill itself. Just as the gusts of wind ripple across the field of barley, so they may pass to one side of either the anemometer or the mill. While smaller windmills were more economical to build, it was thought they could not be coupled together rigidly to form a combined unit owing to uneven reception of the wind.[29] In North America from about the time of the First World War, the Twin Wheel Windmill Manufacturing Company sold a design with two 3.05 m (10 ft.) diameter steel rotors mounted side by side so that the mill would pump large amounts of water for irrigation. The pumps were connected to the rotors by a chain drive, which frequently broke because a gust of wind would hit one rotor and try to start that turning before the other, so gearing had to be substituted.[30]

In the latest version of the Beaufort scale, Force 3 is given as 4.4 m/sec. (10 m.p.h.) and Force 6, 12.6 m/sec. (28 m.p.h.). These are roughly the limits between which the traditional windmill will work and, with only three steps in the scale, it is too coarse for practical use.[31]

Beaufort Scale	m.p.h.	m/sec.	Pressure (lb/sq.ft.)	Old Dutch watermill	English windmill
2 Gentle wind	5	2.24	0.125		
3 Breeze	8	3.58	0.33		will start
	10	4.77	0.5		run slowly
	12	5.36		will start	
4 Good wind	13	5.81			
	15	6.71	1.25		run steadily
5 Strong wind	18	8.05		working well	
	20	8.94	2.0		gives best service
	23	10.28		reef sails	
6 Very strong wind	25	11.18	3.125		maximum useful effect
7 Fresh gale	27	12.07		sails furled	
8 Gale	30	13.41	4.5		sails partly furled
9 Heavy gale	40	17.89	8.0		must be put out of action[32]

The air has mass, which is low because the density of air is low. When this mass (m) moves, the resulting wind has kinetic energy which is proportional to $\frac{1}{2}mV^2$. If then ρ equals the mass per unit volume of air, or its density, and A the area through which the wind is passing, the mass of air passing in unit time is ρAV and the kinetic energy becomes $\frac{1}{2}\rho AV^3$. $\frac{1}{2}\rho AV^3$ is the total power available but A. Betz showed in 1927 that the amount which an ideal wind-motor could extract was 16/27 or 0.593 of the maximum.[33] The actual amount is much less and in 1895 J.A. Griffiths found the highest efficiency in his experiments was 25 per cent in a 3.13 m/sec. (7 m.p.h.) wind.[34] The equations show the increase in energy available in stronger winds because the potential is increasing by the cube of the velocity. At 2.24 m/sec. (5 m.p.h.) the theoretical output of Herne Mill in Kent would be only 1.79 h.p. but at 17.88 m/sec. (40 m.p.h.) a massive 919.98 h.p.[35] In the case of an average traditional windmill, a wind speed of 6 m/sec. (13.6 m.p.h.) will yield about 4.27 h.p. but at 4.77 m/sec. (10 m.p.h.) this will fall to 1.68 h.p. At 3.58 m/sec. (8 m.p.h.), there will be less than one horsepower which, while it might overcome the static friction, would be insufficient for grinding much corn.[36]

Another problem with windmills was getting them to start in light winds. With traditional mills, and indeed most pumping mills today, the machinery which is to be driven has to be put in gear before the mill can start working. The wind must be strong enough to overcome the inertia and the friction of gears and bearings. A modern windpump might run and pump water with a wind speed of 4 m/sec. (9 m.p.h.) but, to start it, a gust of 5 m/sec. (11.2 m.p.h.) may be necessary. The mill will not start without this extra gust. In the same way, the inertia of the mill may carry it through a calmer period of short duration. If the wind drops to 3 m/sec. (6.8 m.p.h.) or lower, the windpump will stop, but if this is only for a brief period, the mill may keep turning until a stronger gust comes.[37] There is no doubt that all the heavy shafting and millwork in a traditional windmill helped to keep it running more smoothly; but this does mean that it is very difficult to harmonise any wind theory with actual practice.

The low density of the wind means that any device to harness it has to be quite substantial to develop much power. This, together with the unreliability of the wind, has meant that more dependable power sources have been preferred even though the energy from wind is free. The history of windpower shows how windmills have been improved over the centuries to overcome these inherent disadvantages. Windmills have been adapted both to generate power in the most appropriate ways for the machinery they have to drive, and also to harness as much as possible of the wind energy available whether the wind is blowing a severe gale or a gentle zephyr. The story of how people have obtained power from wind shows much perseverence as well as great ingenuity.

CHAPTER 2

The horizontal windmill

There are no indications that windmills were ever used for any commercial application in classical Greece and Rome. Hero of Alexandria, who lived around AD 60, described in his *Pneumatica* a windmill which drove bellows for an organ. But the Greek word which has been translated as 'windmill' is used in this sense only in this passage, for elsewhere it seems to refer to a windy headland.[1] The simplest form might have consisted of a horizontal axle with four blades along its length so that, if the lower part were mounted in a box or placed behind a shield, the wind could blow against those sails which protruded over the top. In the later part of the nineteenth century, 'Jumbo' mills like this were built by settlers on the Plains of the United States of America to provide a cheap source of power for pumping water. Sketches in Greek manuscripts show Hero's device without any shield, which must indicate that the wind blew along the axis of the windshaft as in later western windmills. Trips or cams on the axle worked the bellows. This machine remained little more than a toy and led to no further developments at the time.[2] The principle of the wind blowing across the axle of the rotor onto blades set in line with it or radiating out from it is the basis of the horizontal windmill, in which the rotating shaft is vertical so the rotor turns in the horizontal plane. A complete book could be devoted to this subject; but horizontal mills always have been on the periphery of windmill design and will be treated here briefly.[3]

The Persian mill

According to a story of 'Ali al-Taburi (*c.* AD 850) and later writers, the second orthodox Caliph, 'Umar ibn al-Khattab, was murdered in AD 644 by a captured Persian technician Abu Lu'lu'a, who claimed to be able to construct mills driven by the power of the wind and was bitter about the taxes he had to pay.[4] Geographers of the tenth century, al-Mas'udi, Ebn Hawkal and Istakhri, all mention the existence of windmills in Seistan.[5] Al-Mas'udi, writing in about AD 950 said

> Segistan is the land of winds and sand. There the wind drives mills and raises water from streams, whereby gardens are irrigated. There is in the world (and God alone knows it) nowhere where more frequent use is made of the winds.[6]

It was the 120 days of wind blowing from the north during mid-June to mid-October which these mills used, so they always faced in that direction.

How Persian windmills raised water is not known but there seem to have been two versions of the mills used for grinding corn. In one, the millstones were mounted on top of a vertical shaft in the upper room of the mill building. The wind rotor was in the lower part, which was described in the fourteenth century as having

> four loop holes . . . like the loop holes in walls but reversed; as their wider part is outside, and their narrower part directed inward. A channel is formed through which the air-current penetrates powerfully, as through the goldsmith's bellows. The wide end is at the opening and the narrow end within, that it may thereby be more adapted to favour the passage of the air-current which must penetrate the mill building no matter in what direction the wind is blowing.[7]

One of the four 'loop-holes' pointed to each quarter from which the wind might come. Inside was the rotor which might have eight or a dozen sails covered with cloth. The stones were mounted on an upper extension of the windshaft so this form of mill was similar to horizontal watermills.

Rotary horse or donkey mills for grinding corn have been found in the ashes of Pompeii, which was destroyed in AD 79. While the grinding surfaces of these stones were conical, flat hand querns must have appeared quite soon afterwards. Examples of Roman querns, excavated on Hadrian's Wall at Chesters, have grooves incised into their faces to cut rather than crush the grains. It was these horizontal, flat disc stones which were adapted in waterpowered corn mills invented around AD 100. The lower bedstone was stationary and was placed level in framing on one floor of the mill. In the 'underdrift' system, there was a hole in the centre of the bedstone through which the driving spindle for the upper runner stone passed. The runner stone had a larger hole in its centre through which the grain was fed. Across this hole was fixed the 'rynd', an iron bar which rested on the top of the driving spindle and was turned by it. The distance between the stones could be varied by raising or lowering this spindle which regulated the quality of the flour, a process later called 'tentering'. The upper stone had to be balanced very carefully so that it ran true and did not touch the lower stone. One of the tasks of the miller was to re-face the stones regularly, possibly every couple of weeks. In the first type of horizontal windmill, the windshaft from the sails in the lower part of the building must have formed the spindle for driving the runner stone above. Mills like this had disappeared well before 1900.

The second type of Persian corn-grinding mill, which survived until the 1970s or later, had millstones in the lower part of the building and the sails above, necessitating a change to the 'overdrift' system. The spindle, which passed through the centre of the bedstone to a bearing underneath, was still retained to support the weight of both the upper stone and sails and was still used for tentering. The rynd in the upper stone was turned by an iron projection on the end of the windshaft which could be lifted off when the stones had to be dressed. The stones might be 1.5 m (5 ft.) in diameter and their working surfaces were slightly dished. A roughened inner ring, the eye, drew the grain between the stones but the breast in the centre was left smooth to grind the flour, which was expelled by a roughened skirt around the edge. From a hole in the bottom of a hopper mounted on a wall, the grain fell down a chute or 'shoe' into the eye of the runner stone. The shoe was vibrated by a stick which rubbed against a specially roughened part of the upper surface of the runner stone and the rate of feed was regulated by placing wooden pegs in the trough of the shoe. The runner stones rotated at about 30 r.p.m. in a 22.40 m/sec. (50 m.p.h.) wind. Such a mill, working with intermittent wind for about four months of the year, would grind enough flour for about fifteen families.

Mills near Seistan and also on the border between Birtand and Zehedan had six or eight sails interlaced with fresh straw after each harvest. The advantage of having the sails above the millstones was that the sail area could be greatly enlarged. Ropes secured the sails to the windshaft to form a rotor approximately 5 m (16.4 ft.) high by 3 m (9.8 ft.) in diameter. Sometimes matting screens were erected to help channel the wind to the sails.[8] There were no brakes but screens were placed across the wind slots to regulate the wind and, to secure the mill when out of use, the rotor and upper millstone were lowered to rest on the bedstone.[9] The mills were built in a line with high walls separating them and forming shields to direct the wind into the rotors. The famous example at Neh had a long line of 75 mills. At Meshed, Iran, and Herat, Afghanistan, as well as elsewhere in those regions, there were similar mills. While individual mills had to be rebuilt frequently, the long life of this basic design has been quite remarkable, but it could be used only where there was a very strong wind which could be depended upon at certain times of the year.

The spread of windmills

Details about these horizontal mills are found in Arabic sources around the tenth century, which indicates that knowledge of windmills must have existed in the Middle East at that time; such mills were very different from the vertical type found later in the West. It has been claimed that the Crusaders introduced the windmill into Europe. In 1094, the Emperor of Constantinople, Alexius I Comnenus, sent envoys to Pope Urban II asking for help to fight the Infidel. In November 1095, the main features of what was to become the First Crusade must have been worked out, but Jerusalem itself was conquered only in July 1099.[10] The oldest report of a windmill in Israel dates from 1190 and concerns one the Crusaders carried with them to use during the siege of Acre, which is likely to have been of the Western type.[11] With the first windmills making their appearance in England in the years around 1150, it seems unlikely that the Persian horizontal mills had any direct influence on the design of western mills, if only because they needed much stronger winds to work them.

In Tibet, prayer wheels still can be found being turned by the wind, but the origins of this invention are obscure. Again, this region has been suggested as the source of windpower but the lack of any definite early evidence makes this unlikely. The claim that windmills were seen by Fa-Hsien in Central Asia in AD 400 is based on a mistranslation. Prayer cylinders designed for automatic repetition of the famous mantra are unlikely to have been produced before the reign of K'ri-srong-Ide-brtsan in AD 755 to 797 when Buddhism conquered Tibet.[12] Revolving bookcases are recorded by AD 823, but these would have been turned by hand. Early in the twelfth century, a new fashion for mechanical piety swept China; again, it seems doubtful whether this included wind prayer wheels.[13] The use of flags or pennants fluttering in the wind to attract the attention of the gods is certainly very ancient, probably much older than the prayer wheel, but silence about the wind turning prayer wheels suggests that the windmills of Seistan came first.

More is known about the adoption of the windmill in China. Once again early dates for this have been suggested, but these are doubtful. What is probably the true story has been recounted by Joseph Needham. In about AD 1230, Yehlu Ch hu-Tsai was captured by Chinghi Khan and became his minister. He was an extremely good scholar, administrator and mathematician. In his memoirs has been discovered an accurate description of the Persian windmill, with a comment on how good it would be if the Chinese would use it.[14] A Chinese book of the seventeenth century, the *Chu Ch'i T'u Shu*, describes the windmill as if it were a European invention, which could be a mistake for Persia.[15] In China, such mills were used for raising water and evolved into an entirely different form.

The windmills of Persia, Tibet and China, although all having vertical shafts and sails revolving in the horizontal plane, work in different ways and in them can be seen the three basic principles of horizontal mills. These are the shrouded type, the streamlined type and the feathering type. At Seistan we have seen how the wind was diverted through 'loop holes' onto the sails on one side of the rotor where the wind could act on the sails at that side. The masonry formed a pillar which screened the other side of the rotor where the sails advanced towards the wind in the shelter. In this way fewer than half the blades on the rotor would produce full power.

The most common form of wind-driven Tibetan prayer wheel had sails shaped so that they caught the wind on one side of the rotor, but were smoothed or curved to present less resistance on the other. The normal form of blade was a cup or curved sheet so that the wind would be caught in the concave part during half the circle to turn the rotor, while in the other half, the convex or streamlined side would be advancing into the wind and so would present less resistance. The power was derived from the wind pushing against the hollow rather than from any area of low pressure behind it owing to the shape. Power output was minimal although obviously sufficient to turn the Tibetan prayer wheels. This form of horizontal windmill was developed later into the cup anemometer and rotating advertising boards often seen today outside garages or shops.

Chinese horizontal windmills are still used along the eastern coast of China north of the Yangtze, and in the region of Thangku and Taku near Tientsin, to operate chain pumps through gearing for raising salt water for salt pans or fresh water for irrigation. They show a close connection with nautical practice, having sails similar to those on Chinese junks.[16] The canvas sails can be feathered so that those advancing into the wind present the least surface and obstruction. They are mounted at the ends of radial arms in such a way that their supporting masts can pivot. As on board ship, the mast does not pass through the middle of the sail. The longer driving side is tied to the rotor framework by a piece of rope of such a length that, when the sail is turning with the wind, it is held into the wind, but can rotate out of the wind when advancing against the wind. The angle of the sail can be set by the length of rope to a position in which the wind can do useful work on it for more than 180 degrees; in these mills we can see the beginning of the principle of the rotor being turned by an area of low pressure created behind the sail, as well as the pressure of the wind on its face.

The idea of having sails that are fully open during the working part of the cycle and closed or feathered during the return part is one that has attracted inventors over very many years, and many ingenious mechanisms have been devised. While such mills have the advantage that they can work without having to be turned in order to make the sails face into the wind, certain disadvantages outweigh this in practice. As long as only the frontal pressure is being utilised, the speed of the moving surface of the sail must always be lower than that of the wind and therefore the speed of the rotor will remain low. The wind-catching surfaces move in a circular path either into or away from the wind so that they are not all subjected at the same time to the same wind pressure due to their angle and the screening effect on each other. Therefore the power to weight ratio is very low. In the 1750s, John Smeaton considered that the available power of a horizontal mill was only one eighth to one tenth that of an ordinary windmill, but others considered this estimate too low and suggested one fifth.[17] The theoretical maximum power coefficient for a simple horizontal windmill is one third but in practice it will be much less. If complex mechanical systems are used to keep the sails in the optimum positions, then the problems of additional maintenance as well as higher construction costs must be considered.[18] The basic disadvantages of horizontal windmills were not overcome until very recently, when the deliberate use of the area of low pressure behind the sail was introduced.

The first vertical axle windmill in Europe appears as a scheme in the unpublished notebook of Mariano Jacopo Taccola, dated AD 1438 to 1450.[19] The earliest printed illustration was published by Besson in 1578, with a curved four-bladed rotor under a dome (Figure 4).[20] From that time on, most of the Renaissance books of machinery pictures include them. In 1595, Verantius featured five or six different types which he thought preferable and claimed that the western European type was inconvenient

Figure 4. The earliest known printed illustration of a horizontal windmill by Besson in 1578 operating a rag and chain pump for raising water.

Figure 5. The rather fanciful drawing by W. Blith of a horizontal windmill driving a bucket and chain pump. It is probably the earliest printed English illustration of a horizontal windmill (1652).

> because the Axle placed horizontally, which very often has to turn hither and thither, according to the change of Wind – therefore the whole mill has to turn and revolve very easily, and rest and be supported upon a single hinge. Then in such Mills, the Grindstones are set in the upper part, i.e. in a place contrary to their nature.[21]

Verantius pointed out the great advantage of the horizontal windmill, which did not have to be turned to face the wind. In one design, he allowed the sails to collapse on hinges, rather like flaps or doors, which were held in the operating position by ropes. In another, a rotor with a dozen curved sails under a conical cap drove four sets of millstones through gears. Some sets could be disengaged in light winds.[22] There was no braking mechanism.

Horizontal mills in England

The frontispiece of Walter Blith's *The English Improver Improved* published in 1652 features a plate of a fanciful horizontal mill (Figure 5) alongside another of a drainage mill.[23] The numbers of hopeful British inventors can be judged by the records of patents taken out between 1680 and 1870. Horizontal mills constitute the largest group of improvements suggested for windmills but the numbers given below may not include all of them, because the specifications do not always give clear enough details to identify the invention or may cover various inventions of which the windmill is the least. Out of 94 patents connected with improvements to the sails of windmills or smoke jacks, 55 were concerned with the horizontal type.

Date	Number of horizontal mill patents
1680–1700	2
1701–1750	8
1751–1800	12
1801–1850	14
1851–1870	19

These figures show how interest continued unabated well after the middle of the nineteenth century, but most brought only disappointment.

Stephen Hooper took out four patents between 1776 and 1806. The important one for his horizontal windmills was that in 1777 and three mills were built to his designs. The description in the patent about the windmill is brief, for he says only that it consists of

> a shaft or arbour G which is put in motion by the wind on a number of flyers fastened in an angle to arms on arbour G, within the shutters H and I; these shutters open to an angle, by which means the wind is conveyed to the flyers; the shutters are regulated by the wind to shut or let open in proportion to the strength of the same.[24]

The drawing shows a hexagonal pavilion with conical top and vertical shutters round the sides. No details are given of how they were operated nor of the rotor inside. The later drawing in Rees's *Cyclopaedia* shows that the rotor had vertical blades with fixed inclined boards, rather like a waterwheel on its side (Figure 6). The outer shutters could be pivoted to direct the wind onto the rotor or closed competely to stop the mill. In another drawing, elsewhere, all the shutters were linked to an inner iron ring which could be moved round to open or close them so they operated together. The wind must have passed into the centre of the rotor, for there does not appear to be any drum or casing, and found its way out through the blades and shutters on the far side, although one description says 'The other side is completely defended or shaded by the boarding.'[25] Hooper did this in another patent (to be examined later) but then the rotor was altered. If he did close the shutters on the lee side, this must have generated considerable back pressure and loss of power on his first mills.

Hooper's first patent was also concerned with pumping water and shows a bucket and chain pump driven from a horizontal shaft. It is not known whether

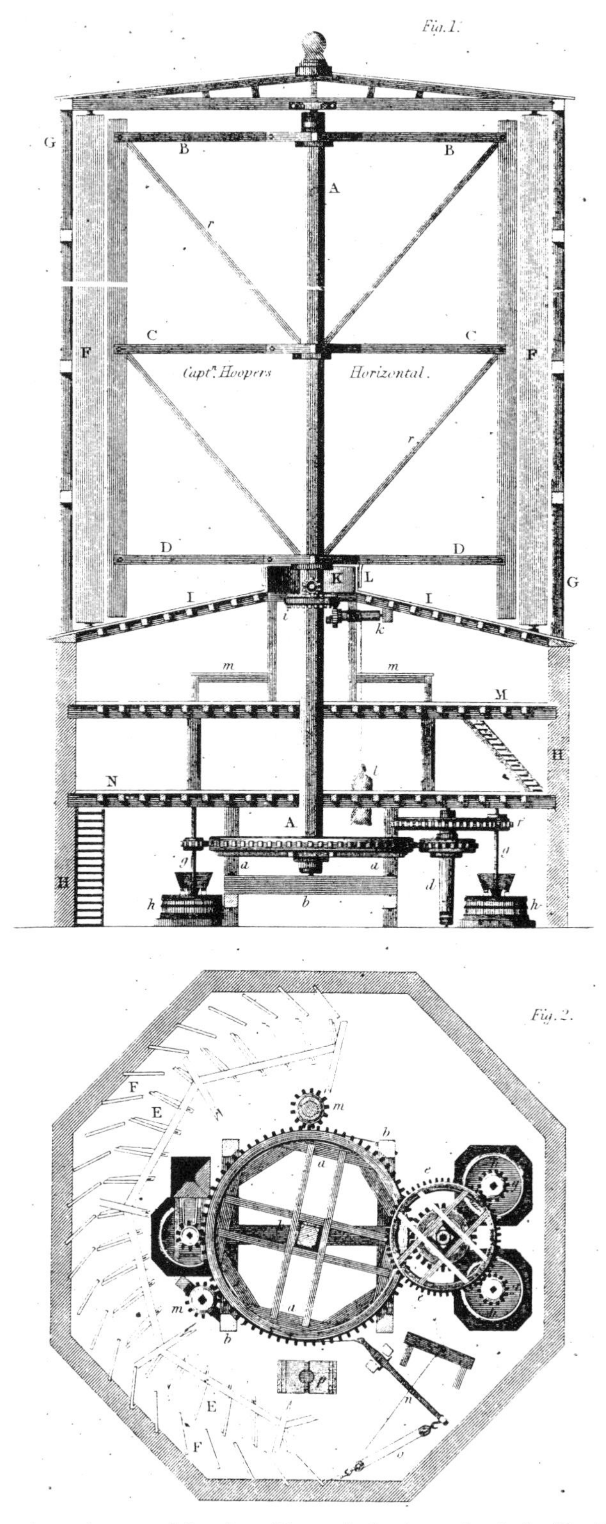

Figure 6. Section through one of Stephen Hooper's horizontal windmills (Rees's *Cyclopaedia*, 1819).

this was the type of pump installed in the Royal Dockyard at Sheerness where Hooper erected one of his mills before 1792.

> The yard is under the inspection of the commissioner of the navy residing at Chatham; from which place the principal supply of water for the use of this garrison, was formerly received; but that expense has within a few years been much reduced by the construction of a deep well, within half a mile of the town. This well is worked by horizontal wheels that will raise, with a smart breeze of wind, two tons of water within the hour; a supply thought sufficient for the wants of the place, and such as has made only one vessel necessary for bringing water to Sheerness, and even that is now considered rather as a job than matter of necessity.[26]

Horses were kept as a stand-by in times of calm.[27]

Hooper built a corn mill on his own property at Margate on top of a two-storey building. The rotor was contained in a tall, slightly tapering tower about 8.53 m (28 ft.) in diameter and 15.7 m (40 ft.) tall. The vertical shaft drove five pairs of stones through gearing. The mill was working in 1825 but had to be taken down soon afterwards after severe storm damage.[28] In 1788, another was built for the maltster, Hodgson, on part of the site of Bolingbroke House at Battersea, then in Surrey. It was designed for grinding linseed but by 1808 was used to grind malt for a distillery. Some figures given for its dimensions, which may not be quite correct, are a total height of 55 m (140 ft.) and diameter of the wheel casing tapering from 21.2 m (54 ft.) at the bottom to 17.7 m (45 ft.) at the top. There were 96 movable shutters 31.5 m (80 ft.) long and 22.86 cm (9 in.) wide which could be operated by a cord rather like Venetian blinds.

> The wind rushing through the openings of these shutters acts with great power upon the sails, and, when it blows fresh, turns the mill with prodigious rapidity; but this may be moderated in an instant, by lessening the apertures between the shutters, which is effected like the entire stopping of the mill, as before observed, by the pulling of a rope.[29]

There were six pairs of stones with provision for a further two. The expense of upkeep was very heavy and, after falling into disrepair, it was taken down in 1849. It was situated close to Battersea church, which gave rise to a story that, when the Emperor of Russia visited England, he took a fancy to that church and had a large packing case made for it. As the inhabitants refused to let the church go, the case remained on the spot.[30] These were the most dramatic horizontal windmills ever erected in Britain.

Many horizontal windmills were produced in America from the 1860s onwards. Among the first were F. and D. Strunk, Janesville, Wisconsin, who patented a design in 1866, and in the same year there was also J.C. Fay, New York. Comparatively large-scale production did not commence until the 1880s and had reached its peak before the turn of the century. Like their annular-vaned counterparts, they were used for pumping water and driving machinery. The structure supporting the rotor was convenient for mounting water storage tanks.[31] In England, Rollason's Wind Motor Company, London, was making horizontal windmills in the 1890s.

Smoke jacks have often been linked with horizontal mills although, in fact, because the wind or smoke blows along their axles and presses on all the blades equally, they ought to be considered as vertical mills. It has been suggested that, because the draught is created by the fire, this is the first form of heat engine and so the first machine to be driven by an artificially produced source of power.[32] Tapering flues above kitchen fires concentrated the flow of gas before it reached the jack. Leonardo da Vinci (1452–1519) sketched a smoke jack with four vanes nestling in the constriction of a chimney. It was linked through a vertical shaft and gearing down to a horizontal shaft from which a rope on a 'V' pulley turned the spit with the joint.[33] Later books of mechanisms such as Zonca show various forms of smoke jacks.

The principle was developed into a form of windmill by Erasmus Darwin, who wrote to the Society of Arts in 1768,

> I have lately constructed the model of an horizontal Wind-mill which appears to have a third more Power than any vertical Wind-mill, whose Sail is of the same diameter, and is in other respects more manageable and less liable to Repair, as it has less wheel-work, having only an upright Shaft on the Top of which the Sail is fix'd. This Model is three feet diameter, and I should not chuse the Expense of sending it up unless there was a *chance*. . . that the Society might grant me a Premium *to assist* me to execute it at large.[34]

A later drawing of this design shows that it had a rotor with six blades like a smoke jack mounted at the top of a long vertical shaft inside a tower (Figure 7). Round the sides of tower were fitted eight banks of horizontal shutters which could be opened on the windy side to deflect the wind up the tower. While the wind pressed equally on all the blades, the direction of the draught had to be changed twice, first as it was deflected upwards in the tower and then when it escaped at the top. Also, with straight edges to the circumference of the blades as depicted there would have been a great loss of power, for the wind could pass through the gap between the blades and the rim of the tower. Josiah Wedgwood was persuaded to install one in 1779 to drive grinding pans for colour pigments at his pottery and sought the help of James Watt, normally associated with steam engines, to improve it. Darwin thought so much of his invention that twenty-one years later he included a drawing of it in his book, *Phytologia*, but it seems that no other examples were built.

After he had moved to London in 1801, Stephen Hooper patented a mill in 1803 which worked on a similar principle to Darwin's. The outer casing was a square, octagonal or similar shape around which banks of either horizontal or vertical shutters were mounted. In this case they were opened on the windward side and definitely shut 'to prevent the wind going out on the leeward side'.[35] The central part of the rotor was 'boarded close' and onto this drum were fitted thirty 'floats or flyers' set at an angle of about 10 degrees and which fitted as close as possible to the rim around the top of the outside shutters. The wind was diverted vertically upwards through the blades around the edge of the rotor. Although Hooper patented improvements to this in 1806, no more was heard of his designs.[36] However, this principle has not entirely died out for it is one of the ideas shown in a recent book on windpower.[37]

The horizontal windmill is far from dead, for development of vertical axis wind turbines has been under way in Britain since the mid-1970s to generate electricity. These rely on the principle of lift with specially shaped aerofoil section blades so that speed of rotation can be increased at the expense of starting torque. The rotor is mounted directly on the top of a vertical torque tube, giving the advantage that all the heavy generating equipment can be placed in the base of the tower where it is easily accessible for maintenance and the tower

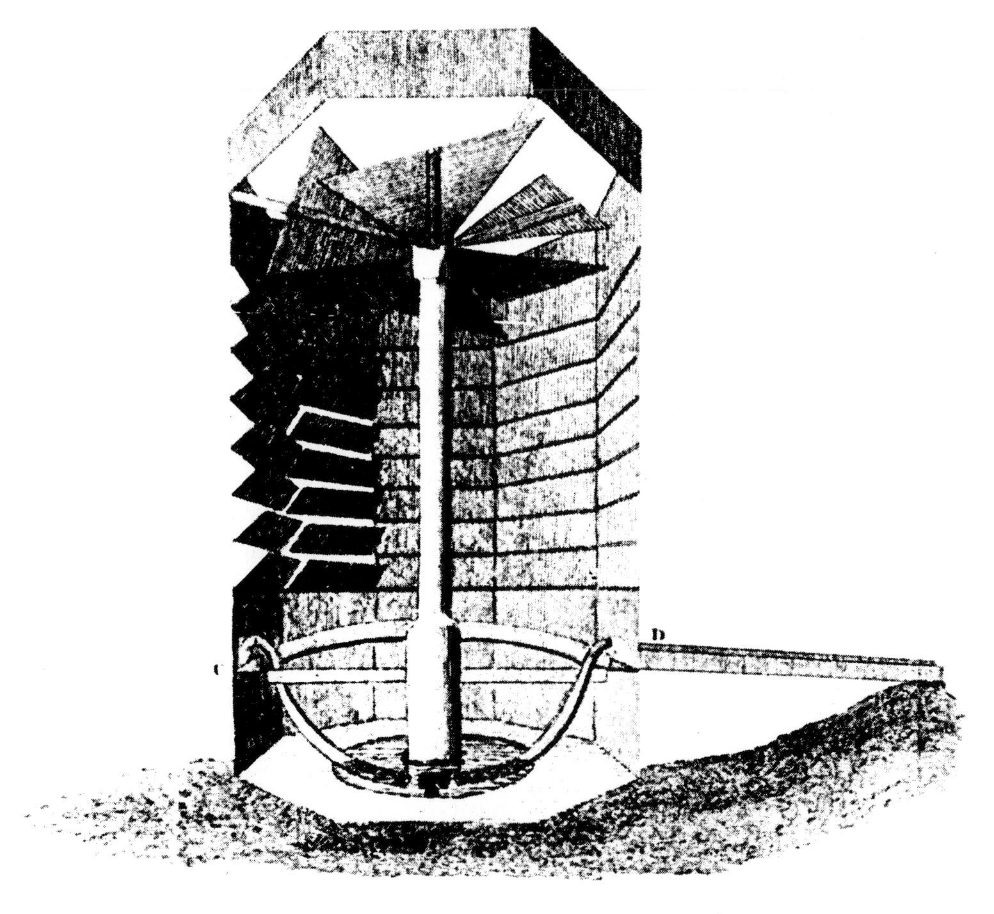

Figure 7. Erasmus Darwin's drawing of his windmill showing the shutters at the left side directing the wind upwards through the rotor (*Phytologia*, London, 1800).

can be made lighter without this weight. One was erected on Sardinia in 1985 and another 17 m (50 ft. 3 in.) in diameter with a rated output of 100 kW in the Scilly Isles in 1987. They have two vertical blades supported at either end of a horizontal cross bar with two sections which can be coned inwards to reduce power output. When the wind speed increases above 13 m/sec. (29 m.p.h.) the angle of the blades inclines to a maximum of 60 degrees on the machine on the Scilly Isles. Here electricity is supplied into the grid to augment either diesel engines on St Mary's itself or the power brought by submarine cable from the mainland.[38] This one is dwarfed by the largest vertical axis wind turbine built at Carmarthen Bay, South Wales, which was commissioned in August 1990.[39] It is too early for performance figures to compare with their 'vertical' rivals.

CHAPTER 3

The post mill

The diet of medieval man would seem meagre and monotonous to us. While improved agricultural methods of that period did help to raise standards of living, the peasant found many of the proteins his body required in the cereals he grew rather than in meat, fish or vegetables. He could not fill up on potatoes or rice because such foods were unknown. In a leper colony in Champagne in the twelfth century, each leper received a daily ration of food of three loaves, a cake and a measure of peas. In 1289 on the manor of Ferring belonging to Battle Abbey, the carters expected cheese in the morning and meat or fish at noon to go with their rye bread and beer. The higher the social category, the lower the percentage of bread consumed. The food budget for different categories of religious and lay workers in the Order of the Hospitallers in Provence in 1338 shows that the monks were allowed forty-three, the lay brothers forty-seven and cowherds sixty-four per cent, respectively, of their annual food expenses on bread.[1] Oats in the form of porridge was the staple diet of many poor people, particularly in the Netherlands. All this grain had to be either crushed or ground. It has been estimated that, in the Middle Ages using the hand quern, one person would have to spend two hours a day grinding flour for the average family; so the benefits of mechanisation are obvious. The trade was mainly local, with the corn being grown in the same area as it was prepared and eaten. People took the grain which they had grown themselves to their own local mill and collected the flour after milling for preparing meals at home. The traditional windmill, with its comparatively small-scale production, fitted into this pattern admirably (Figure 8).

While the Domesday survey of England, carried out between 1080 and 1086, listed some 5,624 watermills in 3,000 locations, there is no mention of any windmill.[2] England technologically was neither ahead of nor behind the Continent and it can be safely assumed that the windmill did not exist. In some areas, watermills were less than a mile apart and over all of England there was an average of one for every fifty households. But there were large areas of the country where there were few streams, or none of sufficient size or fall to power a watermill. The rights to a watermill were owned by the lord of the manor and, in an age when there were no stocks and shares to provide an income the corn mill presented one of the few profitable capital investments available. Therefore, if the lord of the manor could control the provision of flour for his tenants he could derive profit from it, which he was unable to do with hand querns. While small streams were exploited more fully to provide a source of power than may have been the case later after there was an alternative in the form of wind, many villages had only hand querns or horse mills with which to grind their corn. Once the windmill was established, it became another potential capital investment and, with the 'soke' or control over the milling

Figure 8. The post mill at Finchingfield, Essex, dominates the village. The mill is mounted over a roundhouse with the tail pole and ladder on the left. It can be seen how the weather-boarding overlaps the sides to keep the rain out when the mill is facing the wind (May 1990).

rights which the lord of the manor held, the windmill became something worthwhile building and owning, particularly in those areas in the east of Britain where the land was suitable for growing corn but rivers were scarce.

The windmill must have copied the method of milling from watermills for the western windmill is similar in many ways to the Vitruvian or vertical watermill. Here the waterwheel was to one side with a horizontal axle and the drive was transmitted by gearing through a right-angle drive to a vertical shaft, with the stones above in the underdrift layout. There was a type of windmill in Russia which had a similar layout where the horizontal axle below the stones became the windshaft so that the centre of the sails was quite low. A massive timber structure was needed to support the body of the mill. Both the substructure and the body prevented wind passing through the sails so the type must have been fairly ineffective. While the primitive features of these mills such as the underdrift method of driving the stones and the low position of the windshaft suggest an early origin, until more evidence is available, their dating has generally been considered as quite late.

In all other early windmills, the sails were mounted high up to catch the wind. Therefore the windshaft on which they were mounted was above the stones and had to drive them from above in the overdrift position. Presumably early windmills were similar to those about which we do have some knowledge. The diameter of the stones varied according to the size of mill and, while the earliest may not have been much larger than hand querns, later they might range from 1.22 to 1.5 m (4 to 5 ft.). The rynd was at first a form of iron cross rigidly connected to the spindle, allowing no flexibility, but later the 'mace and bar' were introduced. In one version, on top of the spindle was fixed a block of iron, the mace, with a slot across the top in which rested the bar on which the runner stone hung. The stones were housed in a casing, or vat (Figure 9), to prevent the flour from flying all over the mill through the high speed of rotation, for the runner stone might rotate between 100 and 140 r.p.m. although probably in most winds the speed was more like 50 to 60 r.p.m. At Framsden post mill, a sail speed of about 13 r.p.m. with a gear ratio 7.8 to 1 rotated the stones at 100 r.p.m.

The grain was tipped into a hopper mounted on the 'horse' above the casing and the feed regulated by varying the size of the outlet. The shoe dropping the grains into the eye of the runner stone was vibrated through knocking against a squared section, the 'damsel', of the vertical drive shaft. The stones were incised and these grooves or 'furrows' had to be kept sharp by regular dressing (see Figure 14, below). The grains passed into the 'furrows' where they were ground by the higher parts, the 'lands'. The centrifugal force of the runner stone sent the flour to the circumference where it fell down a chute inside the casing to a sack on the floor below.

The theory that the Persian horizontal windmill was turned into the vertical

Figure 9. Overdrift set of millstones with a quant driving the runner. Grain is fed into the hopper down the chute from where it falls into the shoe which is agitated by the damsel on the quant and goes into the central eye of the runner stone. The bell warns the miller when the hopper is nearly empty. On the casing (vat or tun) is a bill for dressing the stones (Bardwell Mill, Suffolk, May 1990).

type and then transmitted by traders from the eastern end of the Mediterranean to the Black Sea, up through Russia and then along the Baltic has been generally discounted, although this possibility has been brought to the fore again in recent research.[3] The principles of the two types are so very different that it seems unlikely that one was converted into the other. That the wind could exert pressure against a surface was well demonstrated in the ships of the Domesday period. Most of those in the English Channel had square-rigged sails which could not sail close to the wind. The lateen sail appears in Greek miniatures of the ninth century and from this was developed the fore and aft rig, on the Atlantic seaboard. The hinged rudder appeared in the twelfth century and this, together with the improved rig, enabled ships to tack against the wind. While the lateen sail spread quickly across the Mediterranean, there is nothing to indicate that ships in northern Europe had progressed to anything more than a single central mast and a single square sail until at least AD 1200.[4] Therefore the lateen sail is an unlikely source for the origin of the vertical windmill.

While the lateen sail draws the ship along through the wind passing across it and producing an area of low pressure behind it, early windmill sails relied almost entirely on the frontal pressure of the wind against their faces to generate power. The wind moves parallel to the windshaft, not at right angles to it, and hits all the sails at the same time for they are mounted like a propellor at one end of the shaft and rotate in the vertical plane. In calculations of the power being generated, only frontal pressure was taken into consideration until well after 1900. If sails on a horizontal windshaft were mounted at an angle to it, the wind would push them both backwards, which had to be resisted by thrust bearings or collars, and also sideways, so they rotated. What the very first windmills looked like is unknown but pictures of medieval windmills show sails made with a light wooden framework over which canvas cloth was stretched to form the surface on which the wind could bear. With nautical experience close at hand, this is the most likely material and form to have been used. These pictures all show mills with four sails, which became the usual number on the traditional western windmill. The genius who invented the windmill, and even the place, remain unknown.

On a watermill, the mill wheel could be started and stopped quite easily by opening or shutting a sluicegate to control the flow of water. During running, the sluicegate could also be used to regulate the wheel according to the speed or load required. The windmill had no method of controlling the strength of the wind and had to make the best use of whatever blew. Canvas had to be stretched out over each sail before starting work (Figure 10). This lengthy task meant that each sail had to be brought into such a position, usually with its tip at ground level, that the miller could unfurl his canvas and tie it over the wooden framework. He had to clothe all four sails, which meant bringing each one round to the vertical position and stopping it there. At the end of the

Figure 10. These four pictures of clothing the last sail on the small Marendijk Wipmolen in Leiden help to show how long it took to prepare a mill for work. It will soon be ready to start pumping with the scoopwheel at its side (March 1992).

working day, the procedure had to be reversed. If the wind rose and became too strong while the mill was working, the sails had to be reefed by the same procedure.

At some stage in the development of the western windmill, a brake system was fitted to the windshaft for stopping the sails turning, but there is no way of knowing whether such a device was employed at first. The sails of a mill can be brought to a halt by turning the mill out of the wind. If the wind no longer blows parallel to the windshaft and directly onto the faces of the sails, the mill loses power. Further round, the mill will stop but if it is turned so far round that the wind blows onto the back of the sails, it becomes 'tail-winded'. Because a windmill is designed to take the force of the wind against the sails from the front, if the wind comes from the opposite direction the windshaft may be lifted off its bearings and the sails blown out of the mill.

There were primitive small mills called 'Chandeliers' in Brittany, France, where the mill was turned out of the wind to stop the sails because they had no brake. The lower part consisted of a round masonry base on which pivoted a circular body containing the machinery. The mill was 'quartered', pushed round so the windshaft was at 90 degrees to the wind by a tail pole. The sails could be spragged to the ground with a stout forked stick to stop them moving and the cloth altered. An example of one of these mills, the Moulin Guezenne at Theolin, was about 6.09 m (20 ft.) in overall height with the masonry base 1.57 m (5 ft. 2 in.) high and 2.51 m (8 ft. 3 in.) in diameter. A central post 30.48 cm (12 in.) in diameter was embedded in the middle and a bearing surface round the top of the masonry was formed with an iron band.[5] In Finistère, a tower mill, the Moulin de Kerteodin, near Cléden Cap-Sizun, was built as late as 1756 without a brake.[6] Mills with primitive sails will not run on bare poles, so once the sail cloths have been taken off, they will stop. However, to stop a mill without a brake must have required turning it out of the wind and judging so it stopped with one sail in reach and furling that canvas. The next sail somehow had to be brought within reach, either by pushing and pulling with a pole or by turning the mill back into the wind, starting and stopping it, furling that sail, and repeating this procedure for all four sails. This must have been quite a performance; it is surprising that a mill should have been built as late as the eighteenth century without a brake.

Brakes were fitted around the main gearwheel or brakewheel that transmitted the power from the horizontal windshaft to the vertical shaft driving the millstones. The principle, that of the externally contracting band brake, is the same across Europe, which suggests a very early origin for its invention. Ramelli gives a description of it in his book published in 1588, but its invention must have been much earlier.[7] To the mill framing or part of the framing in the cap is fixed a link which forms one end of a chain of six or eight wooden blocks surrounding the main brakewheel (Figure 11). The outside of the brakewheel

Figure 11. Inside Bourn post mill, Cambridgeshire. The gearing has been modernised with iron teeth on the brakewheel but the clasp arm construction around the windshaft has been retained. The windshaft passes out of the picture middle left. The ends of the band brake terminate in iron bars, the lower one to a fixed point while the upper end passes down to the brake lever. The vertical quant with stone nut and damsel disappear into the casing over the stones (January 1974).

is one friction surface and the blocks form another. The other end of the chain of blocks is attached to a pivoted lever. Raising or lowering the lever frees or applies the brake. The links and blocks are arranged so that the rotation of the windshaft helps to pull the brake on. The lever is normally sufficiently heavy to put the brake on although sometimes it may have additional weights. A rope passing through a pulley system may be hung down the mill so that the miller can apply or disengage the brake either from inside or from the ground when he is attending to the sails. In some nineteenth century mills, or those which have been modernised, the wooden blocks, which were often made of elm or poplar,[8] have been replaced by a strip of iron running on an iron rim or even a brakewheel made of cast iron. It was the friction of these brakes when the miller tried to stop a mill in a storm which often caused fires and burnt the mill down.

In the western vertical windmill the sails had to be kept facing into the wind and, in effect, created the post mill. The post mill derives its name from the fact that the shell protecting the machinery inside from the elements is turned round with the windshaft on a central pivot, the post. This is the type of mill shown in medieval manuscripts and it could have evolved only in a region where heavy timber was available. The post had to be made from a massive piece of timber to support the whole body of the mill and to stop it blowing over in a gale. On Dale Abbey Mill, Derbyshire, the post was 73.66 cm (29 in.) square at the base and was tapered to 53.34 cm (21 in.) in diameter at the top with a length of 5.18 m (17 ft.). The legend that early post mills could have been built on stumps of trees may have its origins from the lower ends of these posts being buried in the ground. Archaeological excavations have shown that they were not bases of trees because buried also was an elaborate supporting framework. Laid at the bottom were two horizontal cross trees. The bottom of the post was carved into four 'horns' which fitted over the cross trees and prevented the lower end of the post moving sideways. On later mills, the post did not actually rest on the cross trees for it was held above them by quarter bars. These were angled timbers mortised at their lower ends into the cross trees and at their upper ends into the central post. In 1971 at Sydenham Manor, to the northeast of Bridgwater in Somerset, a mound was excavated which was found to contain the cross trees and quarter bars of a medieval post mill dating to the fourteenth or fifteenth centuries. Similar discoveries have been made elsewhere in Somerset and in other parts of the country[9] which show that these 'peg' mills were once common. Just above the place where the quarter bars joined the main post, the post itself was made round so that it could act as a side thrust bearing for the lower part of the body or 'buck'. The bottom floor of the buck was built on two 'shear beams' running from front to back of the mill which formed a 'collar' or bearing round the post. The top of the post carried the whole weight of the buck.

Across the centre of the buck was a large horizontal timber, the 'crown tree', upon which the working part of the post mill depended. First it acted as the upper bearing of the pivot and was carved to fit over a peg or 'pintle' on top of the main post. In later years, this was refined with cast iron bearing surfaces and oiling points. At either end of the crown tree, the whole framing of the mill was secured to form a box structure (Figure 12). Usually 'side girts' were placed on the crown tree running front to back of the mill. Their ends were secured into the middle of the vertical corner posts which were joined top and bottom by the upper and lower side girts. The top front beam, which supported the main or 'neck' bearing of the windshaft, was called the 'breast beam' and its equivalent at the back the 'tail beam'. This main framing was supplemented with additional timbers for strength and also to provide a frame on which weatherboarding could be fixed. Some early mills, such as Bourn and Six Mile

Figure 12. Danzey Green post mill has now been preserved at the Avoncroft Museum. The weather boarding had all fallen off before its removal so that there can be seen the round post supporting the crown tree and at the end of the crown tree the side girts to which the rest of the framing was attached. In this mill, part of the weight of the buck was carried through rollers running on top of the brickwork of the roundhouse (June 1968).

Bottom in Cambridgeshire, had vertical timbers secured to the end of the crown tree and the framing built up on these.

In England the sheathing consisted generally of horizontal overlapping planks to help the rain run off. Those on the front of the mill extended a little beyond the buck on either side to give protection to the joint with those on the sides, while the side planks extended a little beyond the rear wall for the same reason. This worked well as long as the mill faced into the wind but today, when so many mills are preserved but out of use, these overlaps catch the rain when the wind comes from the wrong direction. On the Continent, vertical boarding is the rule in most places. In Holland, the front may be formed with pairs of overlapping planks joined in the middle to make an upside-down 'V' to help the rain run off while the other walls have vertical planks. In the Romanian open air museum at Bucharest, there is a post mill with thatched sides and roof. More usually, though, the roofs were boarded.

Between the top of the crown tree and the front of the mill, a pair of millstones was mounted on framing. In the overdrift layout, there was a direct drive from the stones to the 'stone nut' which was driven by the brakewheel on the windshaft above. The miller could tip the grain into a hopper above the stones while to his right would be the brake lever. Under the stones was the 'bridge tree', a pivoted beam in the middle of which was placed the bearing supporting the vertical spindle carrying the upper stone. By altering the angle of the bridge tree, the miller could alter the gap between the stones and so regulate the quality of the flour. The miller had ready to hand all the controls of his mill.

The heavy weight of the stones, the brakewheel and the sails, all in front of the main post of the mill, were balanced so that it turned easily by the ladder and tail pole. To climb up into the body of the mill, a ladder was built at the back which also helped to support the mill against the force of the wind and keep it steady. In English post mills, a tail pole was secured under the floor of the mill body on the shear beams, passed through the ladder and reached nearly to the ground (see Figure 14, below). To turn the mill into the wind, a lever system on the tail pole, the '*talthur*', lifted up the bottom of the ladder and the miller either pushed the mill round or used a winch on the end of the tail pole to pull the mill round by means of anchor posts sunk into the ground at points around the mill. The tale is told of Scotler Mill in Lincolnshire where a horse was yoked to the tail pole to turn the mill. One day something caused it to shy and turn the mill right round twice before it could be stopped.[10]

A date of 1105 has been claimed for the first windmill in northern France when a 'diploma' was granted by a near relative of William the Conqueror, Count William of Montague, who authorised a Norman abbot to establish windmills in a certain area. However, this document has been proved to be a forgery because it mentions an Abbot of Savigny seven years before the abbey was founded.[11] Dates carved into the wooden framework of mills in Flanders may point to this area being an early centre of windmill construction. In the Moulin de Hofland, a beam has been carved with a much mutilated date of 1114. The Moulin du Nord was moved and extensively rebuilt in 1764, but there has survived a clear-cut inscription indicating 1127 as the construction date. As this was on a side girt and replacement of these parts of the framing is rare, this date could be correct.[12] But the longevity of solid oak beams is well known, so it is conceivable that these parts could have been reused. If these possibilities are dismissed, then the earliest continental references concern three mills in Flanders at Wormhoudt, Silly and Ypres in 1183, 1195 and 1197, respectively. Another at Pontieu, near the mouth of the Somme, is dated to 1191–2, and yet another mentioned in an undated charter at Sainte Martin de Varreville, near Liesville in the Cotentin, has a disputed date of 1180.[13] This gives four or possibly five mills on the Continent before 1200, compared with definitely twenty-three and possibly fifty-six mills in England. The destruction of records and archives in Flanders caused by recurrent wars will prevent further research into possible origins there.

The claim of England for the invention of the windmill has been put forward strongly recently but there are difficulties in interpretation of the evidence.[14] At first, *molendinum* was used for a mill to grind corn powered by a waterwheel or possibly a horse. Once the windmill had appeared, it was described as *molendinum 'de vento', 'venti'* or *'ventricum'*, and then the watermill had the addition of *'ad aquam'* or *'aquaticum'*, but scribes were not always so explicit. We have to remember that, in a power crisis when waterpower was the best source available, even little streams would have driven sets of small stones, possibly no larger than hand querns, which would have been abandoned later after windmills had appeared. Horizontal click mills especially were built on minor streams and have left very few if any remains, such as the many once in the Western Isles of Scotland as well as in Shetland, all but one now gone.[15] In the Ambert region of France, little mills have been seen right up in the hills where waterpower sites would not otherwise have been suspected.

William the Almoner became a monk in 1137 and decided to make a financial contribution by giving his holdings in Wigston Parva, Leicestershire, including its mill, to Reading Abbey.[16] There had been no mention of a mill here in Domesday Book. Sometime before 1200 Ralph of Arraby agreed that the monks should possess 'the whole windmill' (*totum molendinum ad ventum*) and it has been claimed that this windmill was the same as the 1137 mill, making it the earliest in England. But sixty years is a long gap between the definite mention of Ralph's mill and that of 1137 and there could have been a horse mill, or even a watermill for there are small streams in the area. Nearby Wigston Magna had a post mill in 1169 which already had been in the possession of two owners, so it must have been working for some time. These examples show the problems of using early windmill records, for they generally mention a mill only when there is either some dispute or when a donation is being made. Therefore it is most likely that such mills had been in existence for some time prior to reference in a document.

An entry from the *Testa de Nevill* reads

> Hugo de Plaiz gave to the monks of Lewes, the windmill in his manor of Ilford, for the health of the soul of his father.[17]

The document is dated 1155 and refers to a mill in Sussex, possibly above Swanborough Manor. Another mention of a windmill is found in the list of the possessions of the Knights Templar at Weedley in the East Riding of Yorkshire. It was yielding a profit of eight shillings in 1185, which points to an earlier origin. Whether this was the mill included in the promise made by Roger de Mowbray to pay tithes to York Minster thirty years earlier cannot be definitely established.[18] While the site of Weedley village is high on the hill in an ideal situation for a windmill, there are springs in the valley bottom where there might have been an earlier corn mill, and the neighbouring village of Drewton had a watermill in the bottom of the valley in 1100. There is another reference to a mill built since 1180 at Amberley in Sussex by Bishop Seffrid II of Chichester and yet one more at Dinton in Buckinghamshire around 1187. An explosion of references follows, with a further twenty windmills that can be dated confidently to the 1190s.[19]

It is evident that the windmill had appeared sometime before 1155 and that it spread rapidly. References show a quick diffusion of this new technology in the east of England before 1200, stretching from Northumberland right round to Sussex with a significant grouping in East Anglia. We have seen how some had been built on the northwestern seaboard of the Continent, which suggests that the technology was transferred by sea routes. Because the western windmill, with its horizontal windshaft and vertical sails, does not appear outside these locations before 1200, with the possible exceptions of Israel and Portugal, its origins must lie either in England, Flanders or northern France and not in southeast Europe or Asia.[20]

Recognition came in the decretal of Pope Celestine III around 1191. A knight who had built a windmill in his parish refused to pay tithes on it; modern research has shown that the decretal was addressed to the Abbot of Ramsey and the Archdeacon of Ely confirming that the knight should pay up.[21] Windmills also fell under the rules governing feudal rights to grind corn and rivals to existing arrangements might be banned. This happened, probably in 1191, to Dean Herbert's mill at Bury St Edmunds, which he had erected for himself. It enraged Abbot Samson who feared that the income of his own mills might be threatened and the arguments between the two show the problems of establishing a new industry in the medieval world.

> Herbert the Dean set up a windmill on Habardun; and when the abbot heard this, he grew so hot with anger that he would scarcely eat or speak

> a single word. On the morrow, after hearing mass, he ordered the Sacrist to send his carpenters thither without delay, pull everything down, and place the timber under safe custody. Hearing this, the Dean came and said that he had the right to do this on his free fief, and that free benefit of the wind ought not to be denied to any man; he said he also wished to grind his own corn there and not the corn of others, lest perchance he might be thought to do this to the detriment of neighbouring mills. To this the Abbot still angry, made answer: 'I thank you as I should thank you if you had cut off both my feet. By God's face, I will never eat bread till that building be thrown down. You are an old man, and you ought to know that neither the King nor his Justiciar can change or set up anything within the liberties of this town without the assent of the Abbot and the Convent. Why have you then presumed to do such a thing? Nor is this thing done without detriment to my mills, as you assert. For the burgesses will throng to your mill and grind their corn to their heart's content, nor should I have the lawful right to punish them, since they are free men. I would not even allow the Cellarer's mill which was built of late, to stand, had it not been built before I was Abbot. Go away', he said, 'go away; before you reach your house, you shall hear what will be done with your mill'. But the Dean, shrinking in fear from before the face of the Abbot, by the advice of his son Master Stephen, anticipated the servants of the Abbot and caused the mill which he had built to be pulled down by his own servants without delay, so that, when the servants of the Sacrist came, they found nothing left to demolish.[22]

Two points must be noticed in Dean Herbert's arguments. One is that, in Britain, the wind was free. Waterpower was controlled by the riparian owners through which the stream or river passed, hence its easy control by the Lord of the Manor. But in the Netherlands, a tax, the 'wind brief', had to be paid to the sovereign who controlled the rights to use the wind. This tax did not apply to hand or horse mills nor also possibly drainage mills and was abolished by the French along with other feudal dues in 1795. Then, while Dean Herbert claimed that he wanted to grind only his own corn which he no doubt could have done without being stopped, the free burgesses of the town could take their custom wherever they liked and he might have received their custom. Whether the post mill did offer 'quick-witted peasants an opportunity to evade manorial regulation, act independently, and become quite prosperous'[23] is open to question, but it did give yeomen and other free people the possibility of having their corn ground at mills which could be outside the feudal system. The 'soke', or control of milling rights by the Lord of the Manor, continued to govern the development and distribution of windmills where this applied down the centuries. The soke of Liverpool remained in the hands of the Crown until sold by Charles I in 1629, but it was not until 1689 that the corporation took the first steps to abolish these ancient feudal customs.[24] Some privileges continued in other places well into the nineteenth century.

The spread of windmills in the thirteenth century across the eastern parts of England and into northern Europe was impressive, and must be accounted for partly through increasing population. In Europe as a whole, the population is thought to have risen from 42 million in AD 1000 to 73 million in 1300.[25] For Britain, an estimate using Domesday Book would give between 1,250,000 and 1,500,000 for England and from 1,500,000 to 1,800,000 for the whole of England, Scotland and Wales. Growth was very rapid in the hundred years after 1180 and it has been suggested that the population of England and Wales doubled between 1100 and 1300, to possibly 3,000,000. There is evidence of a slight decline before the onset of the Black Death in 1348, when there was a catastrophic fall of perhaps a third, and numbers did not recover again until 1500.[26] In Suffolk, it is thought that, between the Domesday survey and the Black Death, there was a rise from about 70,000 to more than 200,000 people when this area became the most populated part of the country but, around 1380, this had declined to about 130,000.[27]

Both population and windmills spread into the more marginal areas of cultivation, such as the drier uplands or the lower lying swampy lands which could be drained. In this way the windmill supplemented the watermill and did not compete with it. Occasionally a watermill was replaced by a windmill during the thirteenth century, but this was usually where the water resources were unsatisfactory. At Fiskerton in Lincolnshire, a mill which was worth only three shillings in 1125 had been replaced by a windmill by 1231.[28] Yet the greater reliability and steadier running of the watermill meant that waterpower was preferred to windpower whenever possible.

The spread of the windmill can be seen in the estates of the bishopric of Ely, which had fifty manors chiefly in Cambridgeshire but also in Norfolk, Huntingdonshire, Suffolk, Hertfordshire and Essex, regions of good corn-growing country but with low rainfall and slow-moving rivers. In 1086 there were just twenty-two watermills in only twelve of the manors. A survey in 1222 showed only four windmills but by 1251 there were thirty-two, with nine of them being described as new. During this period, the number of watermills had remained steady at twenty. Only at Pulham in Norfolk and Balsham in Cambridgeshire were watermills abandoned in favour of windmills and elsewhere the windmills were erected in manors mostly without mills.[29] In the nearby large estates of the Abbot of Ramsey, which lay principally in the fenlands of Huntingdonshire, it is surprising to find that no windmills were recorded in 1200 in spite of involvement with the decretal of Pope Celestine III ten years earlier. In 1086, only eleven Ramsey manors possessed fourteen watermills, for this is an area with few suitable sites for waterpower. Two Cambridgeshire windmills, at Wistow and Elsworth, could date from 1216 but by 1252 there were still only six windmills. When the next full survey of the estate was made in 1279, there were thirteen watermills but twenty-one windmills.[30]

The slow introduction of the windmill in the first half of the thirteenth century may point to some defect, such as the lack of braking mechanisms, but no description has survived. The spread across the country was uneven with the western parts generally lagging, probably because more waterpower sites were available. In Somerset, a date of 1212 has been given to a reference in the Montacute Chartulary when Robert de Vallibus granted to the church of Montacute, together with his body, his windmill of Sevenamtune with six acres of land close to it. This is the modern Seavington St Michael.[31] In Lancashire, a windmill with the soke of the town of Ince was the subject of a grant to Stanlawe Abbey, Cheshire, in about 1230.[32] The earliest reference to a windmill in Wiltshire is on the manor of Hebelesburnel, identified as Ebbesbourne Wake in the south of the county. In 1248, the mill was valued at one mark. At Sevenhampton, a new windmill was built around 1285 at a cost of £12.2.10.[33] By the 1270s, there were four windmills recorded in Gloucestershire but none in Worcestershire.[34] In Cornwall in 1296, Roger de Langurthou leased to Roger Carpentarius a plot of land on the north of the road which led from Fowey '*versus molindinum venti*'. One was erected on Anglesey in 1303.[35] How many windmills existed around 1300 is difficult to estimate. A total of 10,000 wind and watermills for England has been proposed as a cautious estimate with 12,000 as a possibility.[36] Some of these would have been watermills added onto the 5,624 in the Domesday survey but the remainder would have been windmills and shows the vital contribution they made to grinding the nation's corn and malt.

On the Continent, the early dates for windmills in Flanders and the north of France have been given above. In the thirteenth century, there were 120 windmills operating in the immediate vicinity of Ypres.[37] Dates for their first mills have been given as 1222 for Germany and 1259 for Denmark. The first post mill in Holland was erected at Haarlem in 1274 (Figure 13). This was followed by 's-Gravenzande before 1290, Dessierderkerke near Dordrecht and Amsterdam before 1296 and Koningsveld Abbey near Delft in 1299.[38] Because windmills could operate in freezing winter conditions, they were erected in large numbers on the great plains of northern Europe stretching across Germany, up into Sweden in 1300, Russia and Latvia in 1330 and down into Romania.[39]

The windmill may have appeared at Sienna in Italy in 1237. Dante, who died in 1321, may have seen one for in the *Inferno* he describes Satan, the King of Hell as

> Like a whirling windmill seen afar at twilight,
> or when a mist has risen from the ground –
> just such an engine rose upon my sight
> Stirring up such a wild and bitter wind . . .[40]

In 1332 Bartolomeo Verde applied to the Venetian authorities for permission to erect a windmill; this was granted and he could retain the site for a specified

Figure 13. The Dutch style of post mill. This example with a 'square' roundhouse was seen at Sint Annaland, Tholen Island, in the south west of the Netherlands (August 1990).

period if it should prove successful. On the Iberian peninsula, it has been stated that the earliest windmill to appear 'in unambiguous Iberian records is dated 1182 . . . near Lisbon', although another date suggested is 1261.[41]

At the time of the Third Crusade (1189–92), post mills were exported to the Middle East where they had been previously unknown. It was related, according to Ambroise's eyewitness account, how

> The German soldiers used their skill
> To build the very first windmill
> That Syria had ever known.[42]

How the German soldiers knew about windmills, if their country did not have any until 1222, is not explained. On the walls of the great Templar fortress a little south of Antioch, which was abandoned in 1271, there stood a windmill but its date of erection remains unknown.[43]

As post mills spread across Europe, they were modified and assumed local characteristics at the hand of millwrights. One important development can be traced through the Laws of Oleron, an enterprising commercial community on an island in the Bay of Biscay which promulgated various regulations, some of which were adopted in England around 1314.

> With regard to windmills, some of which are altogether held above the ground, and have a high ladder, and some have their foot fixed in the ground, being, as men say, well affixed; and, accordingly, they are not moveable, for they cannot be detached from the ground, nor removed without damage to their original structure. Of those mills which are actually upon the ground, some people say they are moveable because a man may move them without destroying their original materials; but there is a reason to be contrary. For they are not such machines as tubs, casks, or chests, and still less are they wine-presses, which a man can remove. A windmill is like a house with a ladder, having windows and a fireplace, a cupboard and rooms, and closing with a key and established on its own ground and in its own place. And for that reason it is not moveable.[44]

By this time, the post mill with framing above the ground had appeared. In Sweden, the cross trees were laid on stones on the surface of the ground. Those surviving in England rested on brick or masonry piers to which they were not attached and so gave the appearance of being removable, which did sometimes happen. Not only did raising the timbers protect them from the damp ground but it raised up the whole mill, enabling longer sails or sweeps to be fitted to develop more power.

Post mills were often built on mounds, which at first performed the role of anchoring the buried framing. It has been suggested that this was necessary with the light structures of early mills to keep them steady. Later the mound helped to raise the mill to catch more wind. Sometimes the mound was surrounded by a ditch, possibly to keep out cattle or sheep which might have been hit by the sails. While at first in western Europe the trestles were left open, in northern Europe, Scandinavia and Russia, the mill walls were carried down to quite a low level, sometimes nearly to the ground, to give protection to the framing. In northern Germany, the rear corner posts were jacked up to steady the mill because the ladder did not project backwards but was fixed to the rear and descended alongside the buck. It is not known at what date the 'roundhouse' was introduced to give protection to the trestles. A French illustration of the sixteenth century and a Dutch one of the seventeenth show that the Continent was ahead of Britain, where enclosing the supporting structure does not seem to have happened until the early eighteenth century.[45] The roundhouse, which might be sometimes hexagonal or octagonal, was made weather-tight with a conical roof stretching up to the buck so it protected the quarter bars. This provided a convenient place for storing sacks of grain or flour and had to be provided with a door on either side so that it could be entered whichever way the wind was blowing in order to avoid the sails. However, it was claimed that the roundhouse formed an obstruction to the wind underneath the mill and increased the 'flutter' or loss of propelling force which occurs as the sail passes in front.

Sometimes the cross trees and quarter bars supporting the post were omitted: for example, in the French Chandelier mills where the masonry of the base both located the central post and provided a bearing surface to help support the buck. Romanian windmills had a similar arrangement with a wooden framing across the top of the masonry taking the weight of the mill. In a smaller Romanian mill, in addition to the central post which appears to have been driven into the ground, a bearing frame was constructed on top of four more vertical posts placed around the central one. There were occasions when the main post received additional support from extra quarter bars or sometimes a third cross tree was inserted with its own quarter bars.

As soon as automatic reefing sails had been introduced at the beginning of the nineteenth century and the miller no longer needed to stand on the ground to attend to the cloth on the sails, the roundhouse could be extended upwards to raise the mill into stronger winds. This arrangement was popular in Essex and Suffolk. The roundhouse at Saxtead Green post mill was raised twice so it had three storeys and in the middle one were installed two extra pairs of stones which could be driven by auxiliary power when the wind failed. In each of the floors were four trap doors so that sacks of grain could be hauled up whichever way the buck was facing.[46] Climbing up the ladder into the buck here (see Figure 18, below) or at Framsden Mill, with a similar roundhouse, can be quite an airy experience. In the Midlands, the ring round the top of the roundhouse was sometimes used as a track on which ran rollers supporting the buck to prevent it pitching in strong winds and to lessen the stresses on the main post. In this instance, no roof could be built on the roundhouse so, either the diameter of the roundhouse was quite small to fit under the mill or a flared skirting, such as can still be seen on Wrawby Mill (Figure 14), had to be fitted to keep the rain out. This led into the composite mill where the buck of a post mill was mounted on a roundhouse without the central post, trestle, cross trees and quarter bars, but this construction was very rare.

The oldest post mill surviving in Britain is Bourn Mill in Cambridgeshire

Figure 14. At Wrawby Mill, Humberside, part of the weight of the buck is carried on the roundhouse in the Midland tradition. The tail pole, together with the handle (talthur) to raise the ladder, is at the rear. The millstone in the foreground shows some of the dressing lines (August 1990).

Figure 15. Bourn post mill, Cambridgeshire, is claimed to be the oldest in Britain. The shape of its roof and the small size suggest early origins although a pair of spring sails has been added at some period (January 1974).

which is known to have existed since 1636 and certainly the small size, narrow buck and straight gable roof stem from an early period (Figure 15). The open trestle stands on brick piers which shows that the weight of the mill is sufficient to keep it stable. The buck is 3.12 m. (10 ft. 3 in.) wide by 4.3 m. (14 ft. 6 in.) long with an overall height of 9.46 m. (31 ft. 6 in.). While much has been altered over the years, including two of the sails having spring shutters added and some gearing inside made from iron, the basic layout remains. From the brakewheel on the windshaft, the stone nut on the quant drives the stones directly. It can be put out of gear simply by taking the upper bearing out of its socket or 'glut box'. With such a simple layout, little power was lost in bearings or gearing.

As windmills became larger, more sets of stones could be driven. Possibly when the new sails were added to Bourn Mill, a second gearwheel was fixed to the windshaft from which another set of stones, now removed, could be driven. This 'head and tail' layout became popular in many post mills and probably coincided with the introduction of automatic sails around 1800. The new sails were heavier than the old but the mill could be extended to the rear both to restore the balance and to house extra stones. A different arrangement was to place two sets side by side in the front of the mill. To drive them off the same brakewheel meant complex skew gearing so the power was normally taken through a vertical shaft with a wallower at the top and a main 'great spur wheel' at the bottom driving the stone nuts on either side. At Saxtead Green Mill, the great spur wheel was above the stones so they were overdrift, and this was the most common layout although the underdrift could be found elsewhere.[47] The massive Cross in Hand Mill in Sussex also had a third set of stones behind the post, which was driven by a tail wheel on the windshaft. Add drives to sack hoist and other machinery, and not much space was left inside a post mill for sacks of grain, the bins into which the grain was tipped and the sacks of ground flour.

Lack of space may have been one reason for the development of the hollow post corn mill (Figure 16). The concept must have originated in the Netherlands with the '*wipmolen*' drainage mill. In the hollow post mill, the drive is taken from the brakewheel through the wallower placed centrally in the top of the mill, down a shaft running in the middle of the post which is hollow, to machinery in the equivalent of the roundhouse. In England, there must have been some early drainage mills of the wipmolen type but otherwise here the hollow post mill was rare for grinding corn with the most notable example being that on Wimbledon Common. The type was more common on the Continent. Two survive in the Netherlands at Hazerswoude and Weesp. The 't Haantje at Weesp dates from 1820 and may have been constructed partly from an old wipmolen. While there were some in France,[48] the design seems to have been quite popular in Scandinavia.[49] One advantage was much greater space for the milling processes in the building below. Another was the smaller size of the buck, which contained only the brakewheel and wallower, so that the supporting framework could be lighter. The disadvantage was tending the sails which sometimes must have meant climbing onto the roof of the mill, as in Finland. Some of these mills are found occasionally for industrial use.

The traditional post mill had a very long life. In East Yorkshire, the available

Figure 16. Nieuw Leven hollow post mill at Hazerswoude, Netherlands, shows that little machinery could have been contained in its cap which is similar to those on wipmolen drainage mills. The drive is carried down through a hollow post into the building at the bottom (March 1992).

Figure 17. Chillenden post mill, Kent, although built in 1868, is a typical early design with open trestle except that it is fitted with four spring sails with eliptical springs. The tail pole is supported by a cartwheel for easier turning (April 1991).

Figure 18. Saxtead Green post mill, Suffolk, represented the final development of the post mill with tall roundhouse, fantail on the ladder and patent sails. No working miller would have allowed the trees to grow up and take the wind, but they do now protect the mill from gales (June 1990).

evidence indicates that up until the eighteenth century, all the mills were post mills.[50] In areas with a tradition of millwrighting in wood and supplies of good timber, post mills continued to be built until well into the nineteenth century. In the east of Kent, Chillenden Mill is claimed to be the last post mill to have been built in the country (Figure 17). In 1868, the present mill replaced an earlier one, incorporating parts from its predecessor, and retained the traditional style with an open trestle and post.[51] Wetheringsett Mill, rebuilt in 1881, may have been the last in Suffolk.[52] A few, such as Saxtead Green (Figure 18), Cross in Hand and the Cat and Fiddle mills kept running until after the Second World War as commercial operations, which gives a life span of around eight hundred years for this type of windmill.

CHAPTER 4

Tower and smock mills

Tower and smock mills were built with their machinery inside fixed bodies and normally only the windshaft and sails, mounted in a rotatable cap, turned to face the wind. In a tower mill, the main body or tower would be constructed from any locally available building material such as stone or brick. In England where a hard type of chalk (sometimes called 'clunch') was available, this has been used occasionally, while in Sussex flintstones were sometimes incorporated. In a smock mill, the body was built with wooden framing covered with thatch or weatherboarding. While the smock mill was the lighter construction, the tower mill was sturdier. Sometimes a combination of the two may be found with a smock mill being built on a brick base.

Mediterranean mills

As early as classical times, supplies of heavy timber had been exhausted around the shores of the Mediterranean; this may be one reason why the post mill seems to have been rarely used there, except possibly by the Crusaders. In the Aegean, primitive stone windmills can be found on mountainous ridges in eastern Crete and on the island of Rhodes, where the wind blows from the same quarter for nine months of the year and from the opposite for the other three.[1] These mills were built in the form of a tower with the side exposed to the main wind rounded but the rear against the hill squared. While roofed, they do not have any rotating cap and are called '*monokairos*' because the windshafts always point in the same direction and cannot be turned to face changes in wind direction.[2] In concept, they are the most primitive form of vertical windmill possible but this does not mean that they are the oldest.

Also on Rhodes are some round stone tower mills with parallel sides. This type is common throughout the Mediterranean and the earliest examples of tower mills in northern Europe were similar. Once again, the type is thought to have spread from west to east with dates in southern Europe for Portugal between 1261 and 1325, Spain 1330, Rhodes before 1389 and Gallipoli 1420.[3] They have windshafts mounted in caps which can be turned to face the wind. On Rhodes there is no external means of winding which is done inside by a crowbar: around the top of the tower just below the curb is a series of holes into which a peg can be inserted and used as a fulcrum to lever the cap round.[4] Recent studies have shown that on Crete, windmills were being used during the time when the Venetians were building their fortresses, but whether their origins stretch back before 1300 is open to question.[5]

A woodcut of Rhodes harbour, drawn in 1486, shows tower mills with conical caps and eight sails with parallel blades. This type of sail looks similar to those of western Europe but at some period the Mediterranean cloth sail was introduced (Figure 19); this has been described in modern terminology as a 'full-admission axial-flow' type, but here the similarity to western sails ends for they originated from the rig of local ships. Possibly because timber for heavy western sails would have needed to be imported, the windshaft was extended in front of the sails like a bowsprit, ropes from it helping to stay six or eight radiating spars. The sails were triangular-shaped pieces of canvas like those on a modern yacht. The leading edge of each sail was attached to the spar round which it could be wound to reef it in strong winds. The free corner was secured by a rope so it is as if the wind was forever on the beam. These sails are much nearer to the aerofoil concept and, being flexible, can be trimmed to take up a bigger lift than can the rigid flatter western sails so they have good low-speed torque. The ratio of their tip speed to wind speed is less than one, which means that their low-speed efficiency is high and that they are unlikely to overspeed. The Cretan corn mills had sails about 12 m (39.37 ft.) in diameter with a speed of approximately 25 r.p.m. Twelve or more mills were seen in a row.[6]

This type of sail spread westwards across the Mediterranean and reached Portugal, where pots were sometimes tied into the rigging, which emitted a mournful whistle when working. Presumably the note varied with the speed or change in wind direction and so gave a warning to the miller.[7] Sometimes these sails might have two sets of spars, one behind the other with the cloth stretched between. In the Iberian peninsula, eastern and western types met. In the central highlands of Spain, primitive tower mills survive which have altered little since the first windmills were built in that region some four hundred years ago. Although the stocks for the sails are light-weight and sometimes have to be braced with rigging, the rectangular blades are framed like those further north. Rope rigging would not have lasted long in the wetter climate of northern Europe,[8] but a Mediterranean style mill with eight sails was erected at St Mary's on the Isles of Scilly in 1820 and worked until towards the end of that century.[9]

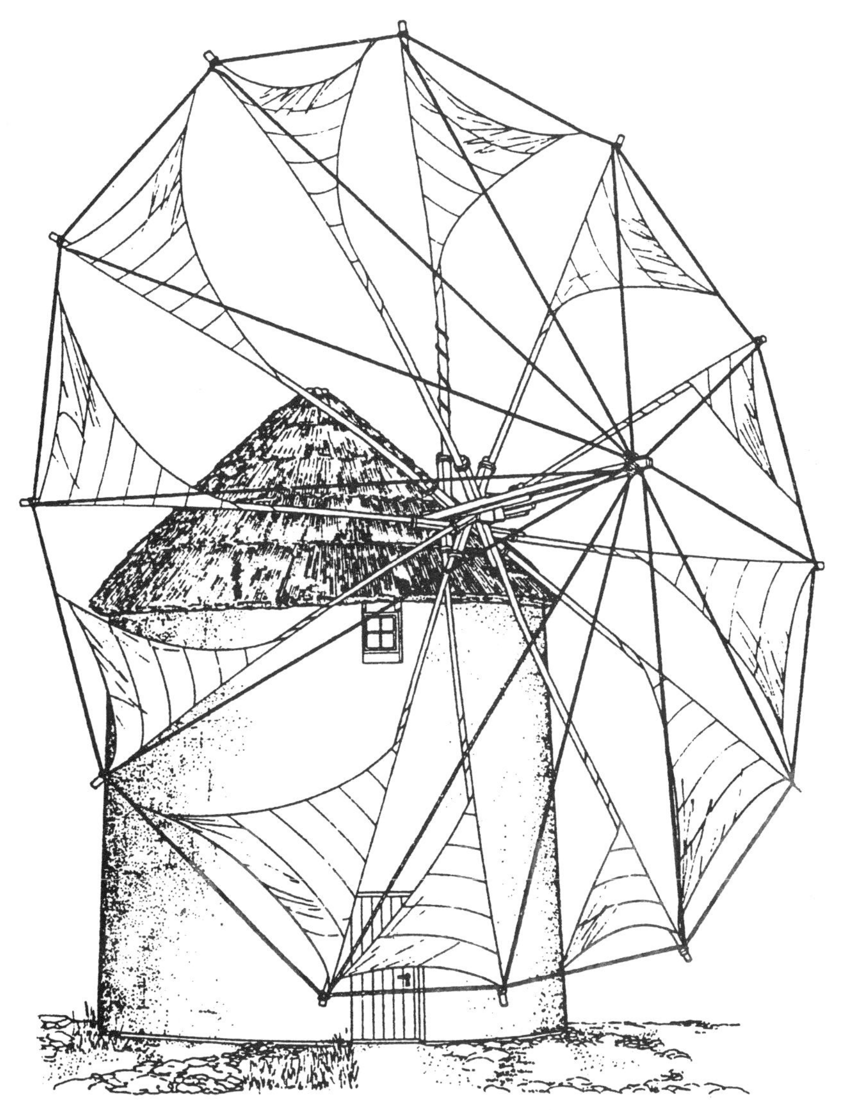

Figure 19. A typical Mediterranean mill with parallel-sided tower, conical cap and cloth sails on a light framework.

Tower mills

The earliest tower mills in northern Europe are thought to have resembled but been earlier than these round Mediterranean mills with parallel sides. Their origins are obscure, with the earliest in France in the twelfth century.[10] In 1295, the accounts of Stephen de Pencastor, Constable of Dover, contain a reference to '*Et in uno molendino ventrico de petra de nuvo construendo in dicto castro*'.[11] Work on making the wall of the mill was begun in that year and cost £36.6.11. A stone windmill of a new construction certainly suggests a tower mill although post mills were erected on the walls of fortified cities where they were well placed to catch the wind, and Carcassonne, France, had some on its walls in 1467. Early maps of towns in the Netherlands show that mills were moved from positions in the fields to the ramparts for better protection from invaders as well as being higher to catch the wind. At Wijk bij Duurstede, close to Utrecht, the brick tower mill, Rijn en Lek, still stands proudly above the town gate by the side of the River Lek. In Leiden, the post mill De Put has been reconstructed recently on one rampart while the famous tower mill De Valk still stands where it was built on another in 1743, replacing an earlier mill. Paris around 1630, as well as Ghent and Copenhagen at the beginning of the eighteenth century, all had mills topping their fortifications. Even London during the Civil War had a windmill surmounting the Mount Mill Fort in 1643.[12]

The first illustration of a tower mill does not appear until 1390,[13] with another in a French Psalter of 1420.[14] There is one of a mill at Cologne, Germany in 1446. The tower mill depicted in a stained glass window at Stoke-by-Clare, Suffolk, dated to around 1470, shows a mill with parallel sides and a conical roof rather in the style of Mediterranean ones.[15] Ramelli's picture of a tower mill, published in 1588, is typical of early ones (Figure 20) and has a circular stone-built tower, tapering in steps up to a conical cap.[16] There is a single tail pole which can be pulled round by a winch and spragged. The windshaft is inclined and the brakewheel engages with a wallower on the top of a shaft driving a single set of stones in the middle of the second floor of the mill.

Six tower mills have been traced in the Netherlands before 1407. Huissen mill certainly existed by 1386 and may have been there since 1373.[17] To the south of Rotterdam, there was a mill at Geervliet in 1382, when people from that place went to Wijk bij Duurstede to buy rollers for their mill cap. This suggests that Wijk already had a mill, though not the one surviving above the gate. A mill existed at Didam in 1408; in 1727 this had parallel outer sides, an exterior staircase and conical cap.[18] The oldest surviving tower mill at Zeddam (Figure 21) was similar with brickwork stepped inside.[19] It was built in 1450 on a mound, which was cut through in 1839 to provide the present entrance, but it has been suggested that a doorway in the side of the tower shows where an outside staircase entered at the level of the meal floor for, originally, there were only two floors. The floor with the stones was above this and opened directly into the cap. The stone chamber is about 7 m (22 ft.) across internally and 9.7 m (30 ft.) externally. It would have contained a single set of stones in the middle driven by a short spindle directly from the brakewheel. Around the top of the brickwork of the tower is a gear ring made of wood (Figure 22). The enormous cap, which is thought to have been conical originally, can be winded by two rope wheels and wooden reduction gearing. The height of the tower does not appear to have been raised and shows that, by the middle of the fifteenth century, windmills had reached the size we know today.

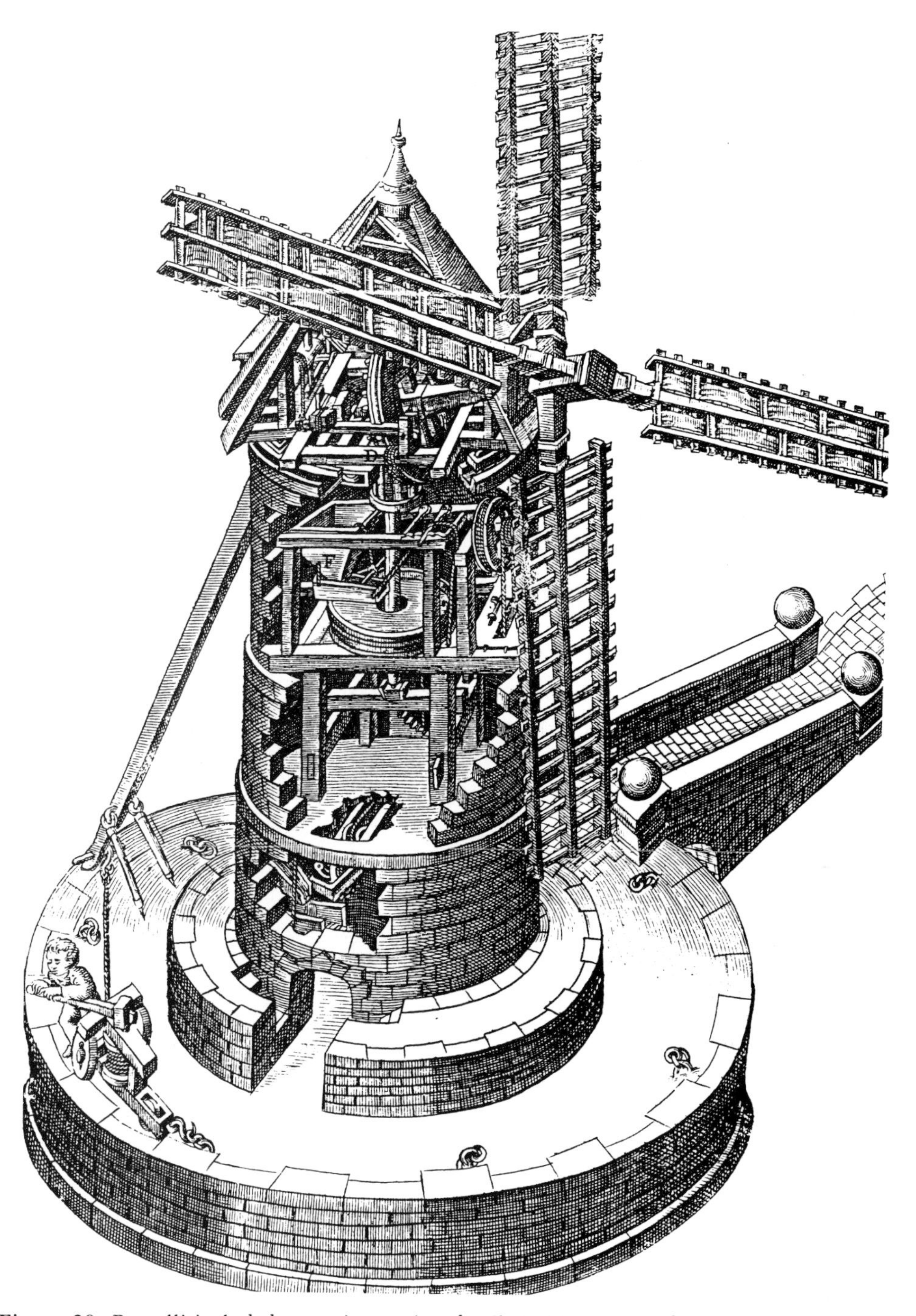

Figure 20. Ramelli included many interesting details in his picture of a tower mill. The cloth is interlaced on the sails and the mill, built on a mound, is winded by a winch. There is only a single set of stones driven directly by the short vertical shaft (1588).

Figure 21. The parallel-sided tower mill at Zeddam is claimed to be the oldest in the Netherlands dating from 1450. It is built into a mound through which the present entrance on the right has been cut. The former entrance was by an outside staircase through the bricked-up lower door on the right. The single set of stones was situated in the top floor (March 1992).

Figure 22. Inside Zeddam tower mill with the wallower on its vertical shaft and the brakewheel behind (March 1992).

The oldest remains of a tower mill in Britain have been claimed for the round stone tower at Burton Dasset in Warwickshire, known to the villagers as 'The Beacon',[20] but the most interesting early survival not far away is Chesterton Windmill to the south of Royal Leamington Spa (Figure 23). The mill is built of stone and six rectangular pillars terminating in rounded arches support the circular milling chamber with its fine windows. These arches would have let the wind pass through the mill so it did not cause so much of an obstruction

Figure 23. The extraordinary design of Chesterton Windmill, Warwickshire, may be partly explained by early tower mills having the machinery on the upper floor with a short driving shaft. This had to be raised high up so that the sails had adequate length (June 1991).

and the same explanation may account for the design of some French tower mills which taper inwards at the bottom. The date of construction of Chesterton was 1632 and there is a suggestion that it was designed by Inigo Jones as an observatory for Sir Edward Peyto, who owned the nearby manor house. There is no documentary evidence to this effect although the building style is typical.[21] Because the upper chamber has straight-sided walls, the cap has to be very large with external diameter about 7.32 m (24 ft.) and internal diameter 4.62 m (15 ft. 6 in.). At some date the curb has been reconstructed, with 15 cast iron rollers running in a cast iron channel to take the great weight of the timbers and lead roof. The mill was winded inside by means of a winch with worm and spur gear which engaged with the beech cogs of a circular rack on the wall within the roller track. To enable the miller to judge the direction of the wind, a weathercock was mounted on top of the mill which was connected to an indicator inside. The span of the sails is about 15.24 m (50 ft.). When the milling machinery was reconstructed in 1776 with two pairs of stones, the original lantern gearing on the windshaft and wallower was retained. Such a layout is similar to the interior of the Dutch mill at Zeddam. Today, Chesterton Mill is preserved with four common sails.

Tower mills with parallel sides were popular in the west of England. In Cornwall, this might indicate links with Brittany or even the Mediterranean. In Somerset, the first records date from the early eighteenth century and refer to mills of modest size and output which may reflect the essentially local trade of that area. Local material was used in their construction, such as the stonework on High Ham Mill, built in 1822 with the upper part of the tower tapering. This is the last remaining mill in the country to feature another local material on the roof of its cap, thatch. Tower mills in this region were generally smaller than those elsewhere and contained only one or two sets of stones. They were about 3.66 m to 4.57 m (12 to 15 ft.) in internal diameter and about 7.62 m (25 ft.) high.[22] Ashton Mill, Chapel Allerton, continued the tradition of parallel-sided stone tower mills (Figure 24). It has been strengthened by iron bands around its circumference, which were fitted in 1900. It has been said that parallel-sided towers were structurally weak and tended to distort under the weight of the cap and sails, causing the curb to sink and jam the cap. The shape of the caps at both these mills once more reflects local tradition.[23] The camera obscura above Brunel's suspension bridge at Clifton in Bristol has been converted from another parallel-sided windmill.

Almost all other tower mills were built in the form of a truncated cone. While the oldest known representation is found in an illuminated manuscript from Normandy dating from 1430 to 1440,[24] the Dutch may have begun extensive use of the type in the sixteenth century, possibly through the influence of the smock mill.[25] At first, the inclination or batter on the walls was not very great so that the size of the caps was not much diminished; but soon a greater slope enable the diameter of the caps to be reduced, so that the weight that had to be carried on the walls was less and the mills could be built lighter. Narrower bodies presented less obstruction to the wind and so gave better performance. The last papermaking windmill, De Schoolmeester, can work in very light winds, due, it is said, to its being built hexagonal instead of octagonal as well

Figure 24. Ashton Mill, Chapel Allerton, Somerset, is a typical small parallel-sided mill. It has been strengthened with iron bands. Winding was by chain and wheel at the rear (June 1991).

as being quite narrow. At first it is probable that these mills were built with the sails reaching ground level. Soon some were built on mounds to catch the wind better and finally multi-floored mills appeared, where a staging was necessary from which the sails could be adjusted.

Tall brick-built mills became common in the Netherlands in the eighteenth century where the highest one at Princenhage was 38 m (125 ft.) to the windshaft with sails 50 m (164 ft.) in diameter.[26] In many towns and villages, they still dominate the sky and one of the most famous is De Valk, The Falcon, in Leiden, built in 1743 (Figure 25). It is 30 m (98 ft.) to the cap, with sails 27 m (88 ft. 6 in.) across. It is said to contain 3,000,000 bricks and has seven floors, the bottom pair forming the miller's house. Because the sails are the common cloth-covered type, they have to be tended from fourth floor level on staging. This is much bigger and stronger than any in Britain because it also forms the anchor points for the winding winch as the tail poles terminated here.[27]

Numbers of tower mills do not seem to have increased much in England until well into the seventeeth century and chiefly in the eighteenth. All the London mills up to 1786 have been described as 'wooden structures turning on posts'.[28] In Yorkshire, the available evidence indicates that up to the eighteenth century all the mills were post mills,[29] while in Lincolnshire it was not until the early nineteenth century that tower mills became common.[30] While a cone was a good shape for taking the wind forces and the thrust against the sails, rainwater did not run off the sloping sides, and this may be why many mills were painted with black tar or covered with some form of rendering. Rain will leach out lime mortar on the side of structures onto which the prevailing winds blow and cause contraction. This must have occurred on windmills, distorting the curb. On many tower mills, windows for the different floors have been placed one above the other. This is said to weaken the building and cracks have been observed running between the windows. On the other hand, the tower mill is the strongest structure out of the three basic types of mill, post, tower and smock. Not only have many tower mills been preserved with their machinery, but, across Britain, the derelict cones of others have been converted into domestic buildings or adapted for other new uses.

The style of tower mills differed in each area. Those in the west of the country tended to be built of stone and, particularly in Anglesey (Figure 26) and Lancashire, had to be strongly constructed to withstand better the buffeting of westerly gales. In the eastern counties, brick-built mills predominated and attained the largest size of any in Britain. Most mills were between 3.66 and 4.27 m (12 and 14 ft.) in diameter at the curb, but Old Buckenham Mill, Norfolk, which was erected in 1819 and was a rare almost parallel-sided brick mill, was 6.7 m (22 ft.) across internally and 12.8 m (42 ft.) high. It also had the largest driving wheels of any in Norfolk, with a 1.6 m (5 ft. 3 in.) diameter wallower and 3.96 m (13 ft.) diameter great spur wheel which drove no less than five sets of stones.[31] The tallest surviving mill is Sutton Mill, also in Norfolk (Figure 27), which is 20.73 m (68 ft.) to the curb (another figure gives nearly 24.38 m (80 ft.)). The present mill probably replaced an earlier one in 1789 but there must have been considerable reconstruction in 1862 after a fire. There are nine floors with four sets of stones on the fifth floor and a later fifth

Figure 25. De Valk, The Falcon, was built on one of the city's ramparts in 1743. The staging formed the anchor points for the winch on the tail poles so it had to be built stoutly. The miller lived in the bottom of the mill. In spite of its size, there are only two sets of stones (May 1983).

Figure 26. The late eighteenth century Llynon Mill, Anglesey, working with only one pair of sails clothed. The blades taper towards their tips. The protruding wooden pegs near the base secure the chain from the winding wheel at the rear of the cap (August 1991).

set on the second. It was struck by lightning in 1940, which put it out of use.[32]

The greatest of all tower mills in Britain was High Mill, Southtown, Great Yarmouth. This was 30.5 m (100 ft.) to the top of the brickwork and 40.54 m (132 ft.) to the weather-vane. Oak piles were driven into the marshy ground and then oak planks laid on top. The base of the brick mill was 910 cm (3 ft.)

thick and 14 m (46 ft.) in diameter. The tower contained eleven or twelve floors and was started in 1812.[33] The total weight of sails with a span of 25.6 m (84 ft.), fantail and cap was around 15 tons; the windshaft weighed 5 tons. In a good wind, it generated 50 h.p. and could grind some 200 quarters of flour weekly. During the Crimean War, in 1854, it supplied flour to Lord Raglan's army; the flour was shipped through the Black Sea in some of Yarmouth's famous clippers. The base of the tower contained an open roadway through which the farmers' carts could pass and unload actually inside the mill.

Figure 27. Sutton Mill, Norfolk, is the tallest surviving in England. The fantail is well clear of the cap to catch the wind. New stocks have been fitted as part of a restoration programme (March 1990).

The mill, which had cost £10,000 to construct, was sold by auction in August 1904 for £100 and soon demolished.

The graceful designs of many Lincolnshire tower mills must rank amongst the most elegant industrial buildings ever. Of these, Trader Mill, Sibsey, may be considered the culmination of the millwrighting tradition (see Figure 54, below). This mill was erected on the site of an old post mill in 1877 by Saundersons, a Louth millwrighting firm, with bricks subsequently tarred.[34] The height is 22.5 m (74 ft.); one third of the way up is an iron staging from which the patent shuttered sails can be tended when necessary. In this case, there are six sails, with a span of 20 m (66 ft.), winded by a fantail with the windshaft concealed in an elegant onion-shaped cap. The framework for the cap is made from iron, as is all the internal gearing, shafting and brakewheel. When a sail was lost in 1950, the mill continued working for some time with the opposite one removed.

The construction of tower mills continued for the greater part of the nineteenth century, when many replaced post mills. Billingford Mill in Norfolk was built in 1860 for £1,300 on the site of an earlier post mill which had been blown over in a gale.[35] In spite of being sited in the bottom of a valley, it was the last corn windmill to work in that county and finally stopped in 1957.[36] It is now preserved, a better fate than Upper Hellesden Mill, Norwich, the last Norfolk corn mill to be built, which was erected in 1875 and demolished in 1920.[37] Probably the last mill to be built in Suffolk was Chattisham Mill in 1867.[38] In Cambridgeshire, the slender brick tower of Stretham Mill was built in 1881 on the site of windmills which had been there since the fifteenth century.[39] While a small tower mill was erected at West Somerton as late as 1899, Saundersons of Louth built the last large corn windmill in England at Much Hadham, Hertfordshire, in 1892. It was 27.43 m (90 ft.) to the top of the brickwork and 35.05 m (115 ft.) to the ball finial of its ogee cap, and was fitted with eight sails. The tower was 12.19 m (40 ft.) in diameter and a railway siding ran into it so that trucks could be unloaded and loaded inside. The mill was demolished in 1910.[40]

Smock mills

The Netherlands is both the home and probable origin of the smock mill (Figure 28). Here western regions lacked good building materials such as stone so timber construction was cheapest. The first examples may not have had any covering over the framing and open, wooden-framed smock mills may have been used as early as 1422 for drainage mills. By 1526 these had developed into 'polder' mills with thatched sides like those surviving today. This type later formed the basis for the industrial mill. Normally there were eight near vertical 'cant' posts which formed the main support for the whole structure and gave the mills their octagonal shape. The cant posts might rest either directly on a wooden piled foundation, on a wooden framework covered with clinker boarding or on a brick foundation which could be a couple of storeys or more high in later mills. This helped prevent rot. The wipmolen was introduced in the south and spread

Figure 28. De Dood smock mill at Zaan, Netherlands, has a weatherboarded base with a thatched top. It was built originally in 1658 and shows the skill of the Dutch millwrights at that period (March 1992).

northwards while the reverse happened with the polder mill. In North Holland, the winding mechanism for the cap is contained inside; this was probably the original method because traces of it can still be found in some mills in South Holland which have been converted later to tail pole winding.[41] Winding by means of tail poles most likely was introduced in the southern part of the country. For polder mills, the traditional covering has been thatch but some Dutch corn mills have weatherboarding.

In England, the octagonal base of Gibralter Mill, Great Bardfield, Suffolk, is thought to have been built for one of the first smock mills around 1680,[42] but the South Lynn oil mill was erected before that by the Dutch in 1638 as a smock mill.[43] A map of 1683 shows at King's Lynn not only this mill but another oil mill near Greyfriars Tower, as well as St Ann's starch mill and Kettle Mill, all smock mills.[44] Today the greatest number of smock mills may be found in Essex and Kent, where the millwrighting tradition in wood survived the longest. Although the cant posts were cut from massive timbers and the intermediate framework was designed to make a rigid structure, buffeting of the wind inevitably caused the joints in the woodwork to move a little. The dyestuff mill, De Kat, dating from 1782 and re-erected on the banks of the Zaan in Holland, can be felt to sway and trembled as it is working. If the prevailing wind is coming constantly from more or less the same direction, the continual pushing and punching from that side tends to force the curb out of the round. This is aggravated by the inclination of the cant posts for, as one side goes down, the other goes up through the post pivoting at the bottom, with the result that one side sinks and the cap may jam.[45] In the recent reconstruction of Sarre Mill, Kent, strong iron plates have been bolted onto the curb to reinforce it (Figure 29).

On the Dutch thatched mills, the thickness of the reed thatch keeps out the rain although, in a strong breeze, the draught may be felt inside. The thatching is continuous all round the mill so there are no weak places; but weatherboarding on an octagonal mill, and this is true if it is either vertical or horizontal, has to join the next side at the corners of the cant posts. On post mills the boarding could be overlapped in a way that prevented the wind driving the rain into the joints but a smock mill body cannot be turned to face only one way into the wind. On one weatherboarded mill at Zaanse Schans, the horizontal planks were observed to overlap each other at the corners but this had to be the same way all round the mill so, while a corner on one side was protected, the opposite corner was fully exposed. This problem results in rot in the cant posts.

Weatherboarded smock mills were much more expensive to maintain than tower mills. Those painted white would need two coats of paint each year and any rotten weatherboards or timber had to be renewed.[46] When most of the smock mills in Kent were built, timber was much cheaper than brickwork. Cranbrook Mill was built in 1814 and still dominates the town (Figure 30) with its 22.86 m (75 ft.) height to the ridge of the cap, comparable with its brick-built contemporaries in Norfolk. The span of the patent shuttered sails is 20.73 m (68 ft.), which originally drove three pairs of stones.[47] The Kent millwrighting tradition in wood continued throughout the nineteenth century, with Stelling Minnis mill being built by T.R. Holman in 1866 and Willesborough Mill three years later by John Hill of Ashford. The final example was built by Holman Bros, at St Margaret's Bay overlooking the Straits of Dover in 1928–9 for generating electricity.[48]

Construction details

On tower and smock mills, the cap is constructed around separate framing which must bear the weight of the sails, windshaft and brakewheel, transmit the wind

Figure 29. During the reconstruction of Sarre Mill, Kent, a strong iron curb is being fitted to prevent distortion. The framing bracing the structure and the interior of the weatherboarding can be seen (April 1991).

Figure 30. The Union Mill, Cranbrook, Kent, is a fine example of a smock mill built in 1814 on a brick base. It was modernised in 1840 with the fantail and patent sweeps (sails) (May 1990).

forces through the main structure to the ground and keep the wind and rain out of the top of the mill. At first the cap was a simple conical structure but distinctive regional shapes appeared. In Holland, there are the thatched caps with windows in the rear. Those in Kent have their ridges almost horizontal but the sides are curved to fit over the brakewheels. In Norfolk, the firm of Dan England of Ludham claimed to have introduced the boat-shaped cap (Figure

31) during the nineteenth century.[49] Lincolnshire mills were characterised by their ogee or onion-shaped caps.

The weight had to be carried on the rim of the tower on a curb, which has been categorised into three types. In the 'dead' curb, the cap turns on blocks fixed to the underside of its framing which run on a wooden track – later an iron track – on top of the curb. In the 'live' curb, rollers or wheels take the place of the blocks. This development must have occurred very early because we have seen how the people from Geervliet in 1382 went to Wijk bij Duurstede to buy rollers. In both of these types of cap, the thrust of the wind against the sails is taken into the top of the tower through horizontal 'truck wheels' attached to the cap frame which run against the vertical side of the curb to locate the cap centrally (Figure 32). In the 'shot' curb, the rollers themselves may help to centre the cap through their shape against the curb. In this instance, rollers may

Figure 31. Looking down on the boat-shaped cap of Hickling Mill, Norfolk, during reconstruction. The canister on the end of the windshaft protrudes at one end while the framework for the fantail balances the other. Part of the brakewheel can be seen in the middle. The cap was lifted into position by a mobile crane (August 1990).

Figure 32. The locating wheels for the cap run on a thrust surface just above the inner ends of the teeth for winding the mill inside West Kingsdown smock mill, Kent. On the right is the bevel gear of the wallower which also has a friction drive for the sack hoist with its chain. A section of the brakewheel can be seen top right (May 1990).

be replaced by ball bearings or wheels with a spherical tread.[50] Both Ramelli and the early Dutch windmill books of the 1730s show that tapered rollers were mounted in the form of a modern bearing with pins through their centres to locate them in a 'roll ring' and keep them at their proper spacing. The weight was taken by the rollers.[51]

The drive from the windshaft had to be taken from the cap into the body of the mill through gears on the brakewheel driving the wallower on the top of a vertical shaft. Here two problems were encountered which were absent from a post mill. As the mill was winded, the brakewheel would move in a circle with the cap. The wallower on its main shaft had to be placed centrally in the mill to keep it constantly in gear. The top bearing of the main shaft had to be located in a heavy beam running across the middle of the cap so it did not obstruct the movement of the brakewheel. The driving forces would be taken against this bearing and beam; when the Dutch drainage mills were being tested in the 1920s to determine their power output, a strain gauge was placed here because the teeth in the two gearwheels tended to separate as power was applied and this could be measured to determine the horsepower.[52] This force will tend to push the windshaft forwards which will be counteracted by the wind. If the windshaft is horizontal, the shaft might move forwards and possibly come out of mesh, which will not happen if it is inclined as in later mills. Because the main shaft will be driving some machinery which will cause resistance to its rotation, the brakewheel could tend to move round the wallower, taking the cap with it. Therefore on a tower mill, whatever mechanism rotates the cap must also prevent it from turning when the mill is working.

Some mills had a lever or some other system inside the top for turning the cap, which was probably the earliest method. At Zeddam, there are now two wooden winches inside the cap which engage with wooden gearing around the curb (Figure 33). In some Dutch drainage mills, particularly those of North Holland, the cap contained a winch with anchor points on the walls. The most general way of turning the cap until the invention of the fantail was by a pole or poles fixed in the cap and projecting away from the sails to the ground. Mills on the Iberian peninsula have but a single pole while the Dutch generally use a group of five terminating near the ground in a main pole fitted with a winch (Figure 34). Around the mill were anchor points to which the rope or chain from the winch could be attached, both to turn the cap and then to secure it. On some British mills, gear teeth were fitted round the outside edge of the curb which engaged with a horizontal worm gear (Figure 35). One end of the worm axle was fitted with a pulley wheel around which passed a continuous loop of rope or chain reaching down to the ground. The miller pulled on the appropriate side of the rope to turn the worm and so wind the mill. A good example with a wooden worm engaging wooden gear teeth can be seen at Bembridge in the Isle of Wight and the principle was applied to the fantail, which will be described later. The advantage was that it was non-reversible, in other words, any attempt to turn the cap would lock the worm and prevent movement. Sometimes ordinary pinion gearing was fitted, which was typical of the mills on Anglesey where Llynon Mill is winded by a chain wheel 2.44 m (8 ft.) in diameter.

The short shaft driving just one set of stones could be turned into a main shaft by placing another gearwheel at the bottom (Figure 36), of sufficient diameter that this 'great spur wheel' could drive simultaneously stone nuts above or below the different sets of stones. Mills equipped like this began to appear at the end of the sixteenth century.[53] Three sets seems to have been the most common, at any rate in Britain in the ninteenth century.[54] The change of the vertical shaft from what might be termed a quant driving a single set of overdrift stones into a transmission shaft from which other power drives could be taken would transform the tower mill into a large milling unit with, in some instances, five sets of stones. Because a tower mill could be constructed both higher and wider than the buck of a post mill, subsidiary machines could be introduced

Figure 33. The inside winding gear for Zeddam tower mill. It is worked by the rope on the left, through reduction gearing to the vertical teeth on the curb (March 1992).

more easily. A friction drive could be fitted onto either the wallower or the vertical drive shaft to operate a sack hoist. To clean and grade the flour, the 'bolter' was introduced in 1502. It consisted of an inclined cylindrical drum covered with a tubular woven bolting cloth. The cloth was rubbed against rods which shook through the finer flour but the bran was retained and dropped out

Figure 34. Conventional tail poles but a modern winch on this Dutch tower mill at Stavenisse, Tholen Island. The chains from the winch are also used to secure the cap to keep it steady (August 1990).

Figure 35. The gearing from the fantail to a worm gear meshing with the ring on the curb can be seen on Thaxted Mill, Essex. The wheel on the right would have worked the shutters on the patent sails (May 1990).

of the end. One of Ramelli's pictures of a watermill contains a bolter.[55] By using cloths with different weaves, the flour could be graded (Figure 37). Later other machines were added too. Also, the tower mill could become the source of power for different industries.

The numbers of windmills

To determine the numbers of windmills is almost impossible. It has been suggested that there were 5,000 windmills in Britain in about 1820 while another estimate gave a peak of 10,000.[56] In 1847 in France there were some 37,000 watermills compared with 8,700 windmills with 90 per cent of the windmills

Figure 36. The massive great spur wheel at Hickling Mill, Norfolk, has a lower drive for other machinery. The bottom of the vertical shaft was located in a bridging box which could be adjusted so the shaft ran true (August 1990).

Figure 37. Flour dresser at Wilton windmill, Wiltshire. A spare cylinder lined with gauze can be seen on the right (September 1990).

being corn mills.[57] Official statistics indicate that, in all likelihood, there have never been more than about 3,000 windmills in Belgium.[58] In Germany in 1875, there were some 54,000 corn mills of which 60 per cent were driven by water, 30 per cent wind (i.e. some 16,200 windmills), and 10 per cent steam.[59] Other figures for Germany give 18,242 windmills in 1895 and 17,000 in 1907.[60] In the Netherlands in 1750 there were between 6,000 and 8,000 windmills, with some 9,000 at work in 1850.[61] Finland reached a total of 20,000 windmills in 1900.[62] Iceland acquired its first windmill only at the end of the eighteenth century, possibly because supplies of good timber there had been exhausted in the fourteenth century and waterpower was readily available.[63]

In Britain, how uneven was the spread is exemplified by the small corn-growing island of Anglesey having over fifty windmills built between *c.* 1730 and *c.* 1835, reaching a total of possibly one hundred windmills, which was the grand total for the whole of Scotland.[64] How closely the windmill was linked with areas of corn production can be seen in Ireland, where many mills were built after legislation was passed in 1784 to encourage the growth of corn there,[65] while some in Anglesey resulted from the demand for corn during the Napoleonic wars. The parish of St Minver, to the east of Padstow in Cornwall, was considered one of the best corn parishes because its land sloped gently down towards the sea. There were many mills here and one was built at Cara as late as 1833.[66] The sixty or so windmills in Cornwall may be compared with the 800 recorded in Lincolnshire.[67] Not all the mills were working at the same time. In Norfolk 949 mills have been recorded, but the earliest comprehensive map of the county, that of William Faden in 1787, shows 256 mills. From the numbers of millers extracted from Norfolk Directories it would seem that between 1845 and 1864 there were over 400 working corn mills with a maximum of about 423 in 1854.[68] This had declined to 350 in 1870.[69] In Suffolk, Joseph Hodgkinson's map of 1776–80 showed 180 windmills, compared with 374 in Andrew Bryant's of about 1823. The first edition of the Ordnance Survey in 1836–8 showed 430.[70] A similar pattern can be observed in Sussex where, during the early part of the eighteenth century, approximately 80 mills were being worked; this had increased to 190 by 1823.[71] The number of windmills in England probably doubled between 1760 and the 1820s, although the proportion of the national horsepower that they provided halved over the same period to approximately 6 per cent.[72]

This continued rise in the number of corn mills must be seen against a background of rising population and also the introduction of the rotative steam engine. Around 1500, the population of England and Wales may have been 3,000,000 and in 1600 possibly 4,500,000. It is a fair estimate that in 1700 the figure was between 5,000,000 and 5,500,000 and, in 1750, about 6,500,000. Between 1750 and 1801, the year of the first Census, there was a growth of 40 per cent to between 9,000,000 and 10,000,000. The most rapid expansion came between 1801 and 1821 when there was an increase to 12,000,000. The next doubling took less than sixty years.[73] The windmill helped to grind the corn to feed these extra mouths and freed scarce water resources for other industries.

However, Thomas Newcomen had built the first successful steam engine in 1712 but at first it could be used only to pump water. John Smeaton, who had improved the performance of Newcomen's atmospheric steam engine, wrote to 'The Honorable the Commissioners of His Majesty's Victualling Office' in May 1781 about the best method of driving a corn mill at H.M. Victualling Yard, Deptford. He could not recommend a steam engine because

> All the fire-engines that I have seen are liable to stoppages, and that so suddenly, that in making a single stroke the machine is capable of passing from almost the full power and motion to a total cessation . . . In the motion of mill-stones grinding corn, such stoppages would have a particular ill effect.[74]

While wind might fail suddenly, the momentum of a windmill would bring it to a gradual halt and not cause damage as, Smeaton anticipated, would the sudden halting of a steam engine. Within five years, in April 1786, one of James Watt's new rotative steam engines was driving six pairs of millstones in the Albion Mill, London, on the Thames by Blackfriars Bridge.[75] He had patented his separate condenser in 1769 which made dramatic improvements in fuel consumption but the steam engine still remained only a pump. Two further patents, that for a double-acting engine in 1782 and his famous parallel motion in 1784, enabled him to develop a steam engine which could drive rotating machinery directly and compete with windmills.[76] The Albion Mill was the very first building purposely designed to use rotary steampower and, in size alone, represented a very great advance on previous wind and water mills. The largest corn mill in London at that time had four pairs of stones, but at the Albion Mill the plan was to install three engines each driving ten pairs of stones. A second engine was erected in 1788, but the whole venture ended in a spectacular fire during the night of 2 March 1791.[77] However, these engines had shown the power of steam and were the precursors of many more steam mills.

CHAPTER 5

The sails

Medieval scribes must have enjoyed drawing pictures of peasants delivering sacks of corn to post mills or millers trying to wind their mills with a dog looking on. These little sketches around the edges of manuscripts were not intended to be accurate representations from which later historians of technology could make reconstructions, so an element of doubt must remain about specific details. While archaeological excavations have exposed foundations of early post mills, their superstructures have rotted away long since so no verification can come from this source. Yet there are some consistent features in pictures of early medieval post mills from which some conclusions may be drawn. All the illustrations show that a set of sails was composed of two pairs of blades, with one of each pair fitted onto either end of a common stock which passed through the end of the windshaft. In modern parlance, they were the 'full admission, axial flow type'.

The first sails

The wooden framing of each blade was covered with cloth presumably made from canvas like that on ship's sails, as it is still today, and was secured in two different ways. On some, the cloth was interlaced through the framing (Figure 38) and this tradition continues through the drawings of Diderot's Encyclopaedia published in 1763,[1] up to the beginning of this century in Cape Cod, U.S.A.[2] In these sails, the stock passes through the middle and the framing extends equally on either side in a single bay so that the sail can be covered with two pieces of cloth which are woven in and out of the cross bars and made fast top and bottom. They were reefed 'by being drawn towards the centre like a curtain'.[3] While this operation would be difficult in a breeze on a summer's day, it must have been almost impossible in the middle of a freezing winter's storm because the canvas could not have been handled around the bars. In the Netherlands,

Figure 38. Ramelli's drawing of a post mill shows sails with no twist, interlaced with cloth (1588).

this system was completely replaced in the seventeenth and eighteenth centuries.

On a pleasant summer's day with a stiff breeze merrily turning the sails spread with full canvas, the windmill is a most attractive and romantic sight; but conditions in winter must have been grim. Because a windmill really needed gale-force winds to work effectively, the miller had to take advantage of whatever winds he had and, in a rising gale with rain or sleet lashing down, he would have to work for as long as he dared before going outside to furl his sails and prevent the mill running away in the strong gusts. To handle sodden, possibly frozen canvas with water pouring off and running down the arms, through the clothing round the body and out at the boots, reminds us of what working conditions could be like. And the cloth had to be altered on each of the four sails in turn, often in the dark.

Common sails

Other pictures show cloth tied on at the corners and some intermediate points on the front of the blade where the pressure of the wind helped to secure it. The power developed depended upon the strength of the wind and the area of the cloth so that, to regulate the mill, the miller had to vary the amount of sail exposed. When out of use, the cloth on these 'common' sails would be rolled up and secured to the framing. The inner end could open out easily because the corners had eyelets to which were attached chains or which could run on a bar so that the miller did not have to climb up as far as that end when he was setting sail. Further down the whole length of either side selvedge were cords and 'pointing lines'. The cords attached the leading edge while the pointing lines ran round cleats on the trailing edge and could be set quickly to alter the area of sail exposed. According to the amount of wind and power required, the sail could be set to 'first reef', 'dagger point' or 'full sail'. The appropriate pointing line was passed round its cleat and secured while the rest of the cloth would remain furled.[4] This sail design spread across most of Europe and has continued in use up to the present day.

In parts of northern Europe like Sweden, Finland and Russia, and into North America, longitudinal slats of wood might replace cloth, being easier to handle in freezing conditions (Figure 39). Boards were made up into sections which in South Prussia were known as 'under' (at the tip), 'middle', 'storm' and 'permanent'.[5] The permanent boards were fastened with nails and covered the inner part of the arms. They provided sufficient surface to drive the mill in high winds, while the other boards could be fixed with wooden pins passed through staples to provide power in moderate winds.[6] These sails with wooden boards, and the medieval illustrations, reveal that early sails had constant weather and were straight without twist. This is confirmed by Ramelli. Smeaton said that such sails were most commonly set at an angle from 72 to 75° with the axis at 15 to 18° to the plane of rotation which, while it might have given a reasonable starting torque, would have been inefficient at any higher speed. During tests in 1895, J.A. Griffiths found that, to start the mill from rest, the best angle of the blade should be from 70 to 55° but on most actual examples he found that the greatest angle was 43°.[7] Once the mill begins to turn, the angle of attack should alter according to the speed of the blade. At higher speeds, the blade needs to be flatter, or turned to a position nearer a right angle to the wind. The tips will be moving much faster than the centre so the angle of the blade should vary along the length. A twisted blade was found to improve performance even though modern research points to there being little difference between flat and twisted blades.[8]

Figure 39. These little post mills on Oland Island, Sweden, would have had boards fitted to their sails when working (June 1978).

Figure 40. The sails on the Fuglevad windmill from Zeeland and now at the Copenhagen Open Air Museum show the twist along their length. This smock mill is covered with wooden shingles (August 1965).

Sometime during the sixteenth century, Dutch millwrights perfected the design of the traditional windmill sail which has remained unaltered up to the present (Figure 40). The best form seems to have been arrived at empirically, presumably through experience without any testing and this new type of sail may have led to the great flowering of the Dutch industrial mill at the end of that century and in the early years of the next. Sails with a double curvature are shown in the patent of Cornelis Dircksz taken out on 31 October, 1589, and it appears on Jan Brueghel's painting of 1614.[9] However, in his calculations of the power of a mill around 1610, Simon Stevin correlated the wind pressure with the width of the sails for flat sails only. He noted whether the sails on which he was carrying out his experiments were old or new designs, which suggests that the improved sails had not been in use for long. The framing and cloth were nearly always entirely on the driving side of the stock or 'whip'. The whip was the long main spar of the blade and might be a separate section which could be bolted onto the stock so that the stock did not have to be the full length of a pair of sails. The leading side might be formed of a narrow 'leading board' set at an angle, generally 22½° to the whip, which helped to guide the wind onto the sail. In the Netherlands, the angle of the leading board differed according the type of mill. Those on paper, saw and polder mills were set at a steeper angle because these mills needed to generate more power.[10] Before any holes were drilled into the whip for the framework forming the driving side, the positions and angles had to be carefully laid out so that the spars would be at the correct inclination to give the weather twist. Sails like this are shown in the famous Dutch windmill books of the eighteenth century.[11]

Theory of power from wind

People like Stevin were soon asking whether this was indeed the best form. Stevin was able to compute the power of his mills by working backwards from the amount of water that they pumped. He could determine this by knowing the number of revolutions made by the scoopwheel, the depth the blades dipped in the water, their width, and the height to which the water was raised. One of the drainage mills in the Beemster polder in 1607 had a scoopwheel 5.18 m (17 ft.) in diameter, 30 cm (1 ft.) wide, with a depth of immersion of 610 cm (2 ft.) and a difference in height of 910 cm (3 ft.). About 80 cu. ft. of water were ejected for each revolution with a normal speed of 9.4 r.p.m. The wind-shaft was turning at 13.5 r.p.m. and, through knowing the gear ratio, it could be calculated that about 10 h.p. was being generated. From this Stevin tried to calculate the minimum wind pressure needed to move his scoopwheel by taking the area of the sails and the pressure on each square foot. But he failed to connect the differing velocities along the length of the sail depending on the speed of rotation, the wind velocity and the energy available at the scoopwheel shaft because he had no way of measuring the speed of the wind.[12] Possibly through these difficulties, Dutch eighteenth century windmill books contain constructional details for building mills but no theory on how they work.

Among people who tried to discover principles for windmill performance was Francis Bacon, in England. He wrote in 1622.

> There is nothing very intricate in the motion of windmills, but yet it is not generally well demonstrated or explained . . . The wind rushing against the machine is compressed by the four sails and compelled to make a passage through the four openings between them. But this confinement it does not willingly submit to; so that it begins as it were to joy the sides of the sails and turn them round, as children's toys are set in motion and turned by the finger.
>
> If the sails were stretched out equally it would be uncertain to which side they would incline. As however, the side which meets the wind throws off the force of the wind to the lower side, and thence through the vacant intervals; . . . But it should be observed that the origin of the motion is not from the first impulse (that which is made in the front), but from the lateral impulse after compression has taken place.[13]

Bacon's recognition of the compression of air or pressure against the blades is significant for it was this force that his successors would use to calculate the power. He experimented with bellows to produce a draught against paper sails of various shapes and found he could use a gentler draught with eight broader sails, but realised that the weight of extra sails might become too great. The length of the sails, too, affected the performance and he wondered whether it would be advantageous to enlarge the tips, but did not experiment with this. He gave the warning that 'If these experiments be put in practice in windmills, the whole machine, especially its foundations, should be strengthened'. He must have realised the difference between theory, models and the full scale, and it was the difficulty of translating small-scale tests into actual practice which for centuries hampered the development of successful theories on windpower.

It is a pity that Sir Isaac Newton did not carry his youthful interest in windmills into his maturity. While a boy, a new mill was erected in Grantham and the lad made a model of it which he put on the top of the house where he lived.[14] Yet, in his *Principia*, there is nothing in his laws of motion which suggests that he took into consideration the performance of windmills. It was left to the French to begin to propose theories of windpower based on experience gained from waterwheels. In 1704, Antoine Parent had suggested that the efficiency of an undershot waterwheel turning at one third of the velocity of the stream was about 15 per cent when compared with a perfect one which ought to be able to pump back again to the original height the water which drove it. Parent realised there must be a connection between an incompressible fluid, water, and a compressible fluid, air (although he did not use these terms), and that it ought to be possible to determine the power of a windmill from the speed of the wind.

The relationship between air and water was further explored by Belidor in the 1730s.[15] Parent published his calculations in 1713 in his *Recherches de Mathématiques et de Physique* and showed that the wind force on the sails was proportional to the square of its speed and the square of the sine of the angle with which it hit the sails. He went on to deduce the best angle for that inclination should be 54° 44', giving 35° as the angle between the plane of the sail and the plane in which the whip rotated when the force being developed by the mill was equal to $^{5}/_{13}$ of the force of the wind.[16] His calculations were to form the basis on which subsequent engineers worked for the next fifty years both on the Continent and in England.

W. Emerson, in England, followed Parent but by the time his book, *The Principles of Mechanics*, was published in 1754, it was realised that Parent had neglected the effect of rotation on the inclination or angle of incidence of the blades. Emerson used the angle of incidence quoted above as the best angle for starting a mill and pointed out that this

> would always be so, if the wind struck them in the same angle when moving as when at rest. But by reason of the swift motion of the sails, especially near the end G [the tip], the wind strikes them under a far less angle; and not only so, but as the motion of the end G is so swift, it may strike them on the backside. Therefore it will be more advantageous to make the angle of incidence WCF greater, and so much more as it is further from E.[17]

This was only putting on paper what the millwrights had already discovered practically, but Emerson advanced no theory or calculations. This was left to the Frenchmen, Léonard Euler and Jean L. d'Alembert who, around the middle of the eighteenth century, continued theoretical work on the angle of the sails but were overtaken by the experiments of John Smeaton.

John Smeaton

It has been claimed that John Smeaton was 'the first investigator of the problem of windmill sail design on scientific lines'.[18] However, in his paper to the Royal Society in 1759, Smeaton himself pointed out that a Mr Rouse, of Harborough, Leicestershire, and Mr B. Robins had experimented earlier with models.[19] He also drew on the earlier papers of Maclaurin and Parent and had looked at windmills on a Continental tour. Smeaton is noted for his theory, based on tests which were carefully planned and executed. Before his windmill experiments, he had built a hydraulic test rig with which he was able to compare two different types of waterwheels under varying conditions of flow with various simulated loads. This worked well and he was able to discover which performed better. But, when he turned to windmills, he ran into difficulties. He recognised correctly that 'the wind itself is too uncertain to answer the purpose, we must therefore have recourse to an artificial wind'.[20] With the equipment available in his day he realised that he could not generate a wind 'in sufficient volume, with steadiness and the requisite velocity'. He also recognised that, if he were to try to create a false wind by moving his test rig in a straight line through the air, he did not have a large enough room. So he placed the sails or rotors to be tested on the outer end of an arm 1.54 m (5 ft. 6 in.) long which could be rotated horizontally, driven by someone pulling on a rope which had been wound round a drum under the arm while watching the swings of a pendulum to regulate the speed (Figure 41). The sails to be tested were of 53.34 cm (21 in.) radius and were mounted on a horizontal axle which also acted as another drum on which the cord from a weight hanging vertically could be wound so that the rotation of the sails drew up the weight. The size of the weight could be changed and the performance was calculated through the length of time it took to raise the weight a certain height.

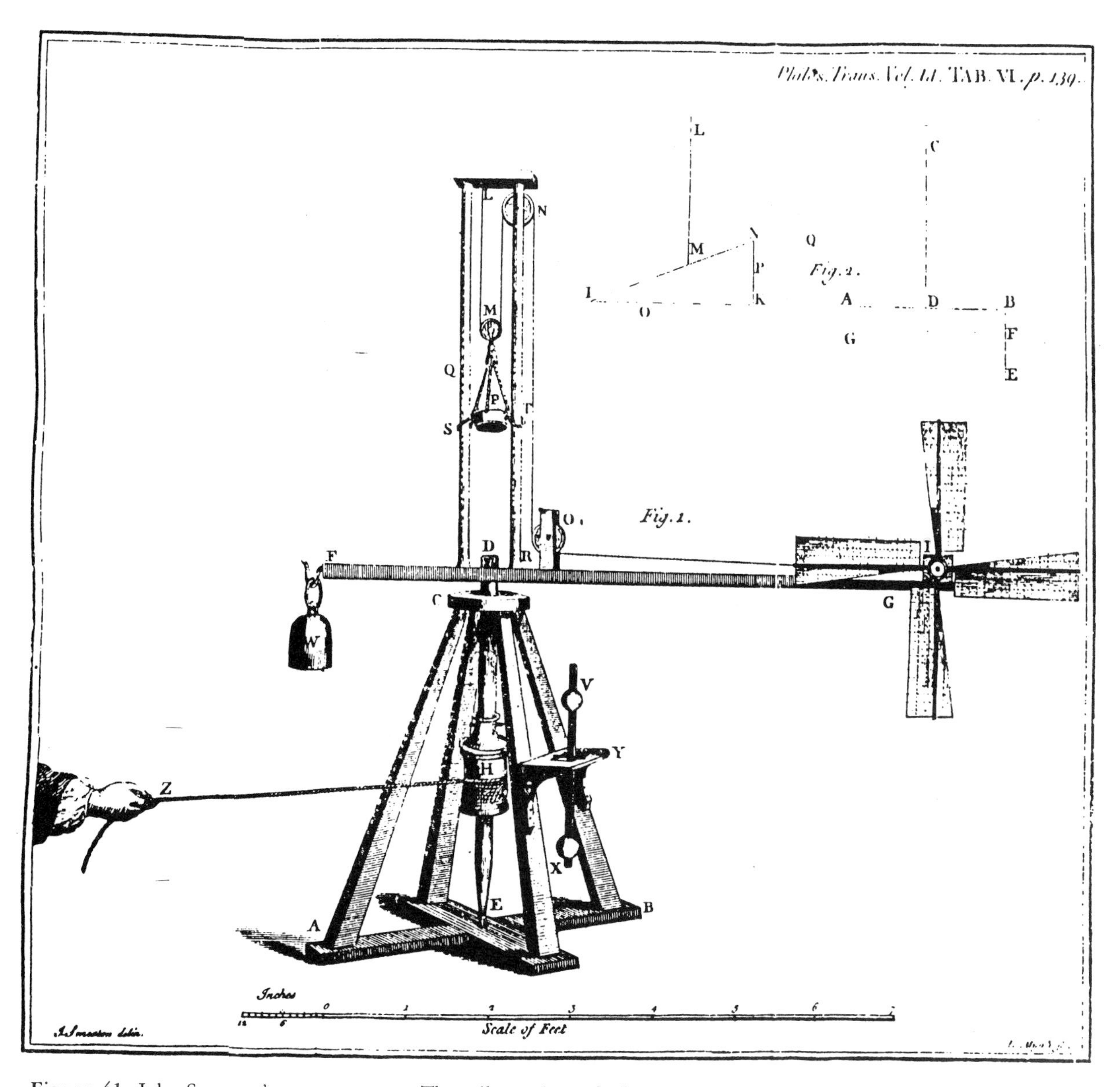

Figure 41. John Smeaton's test apparatus. The sails on the end of their arm were driven by the rope pulled by the hand in time with the pendulum VX. Weights could be placed in the pan P to determine the power developed (1759).

Smeaton carried out nineteen experiments on at least six different types of sails which could be set at different angles of inclination. While the tests were comprehensive, they suffered from the defects of being in an artificial situation and also there was no equivalent for the body of the mill to cause an obstruction to the wind. His illustration shows only one type of sail so we do not know the shape of the others. He took great care to compensate for friction and claimed that his results were borne out in practice because he had had an opportunity of verifying them on an oil mill which he had constructed.[21] He put forward eight 'observations and deductions' which contained a further nine maxims. He pointed out that the theoretical statements by Parent and Maclaurin were wrong. While Parent's value of 35° for the angle of inclination between the plane of the sail and the plane of rotation of the whip gave the greatest force, the speed was too slow for any useful work. Smeaton found that half this angle gave an increase of 50 per cent in the product of the load at the maximum number of revolutions. He determined that the best angle for the weather would decrease from the centre and gave a table worked out in stages of a sixth along the length.

	Angle with the plane of motion	
Successive sixths	Smeaton	Templeton, 1856
1	18°	24°
2	19°	21°
3	18°	18°
4	16°	14°
5	12½°	9°
6	7°	3°[22]

Later figures differed slightly from Smeaton's. His results indicated that the best shape for the sail would be convex on the windward side but Dutch experience had found that a concave form was better. This was due partly to the need for the tip to be nearly parallel to the plane of rotation in order to avoid creating back pressure when the sail passed in front of the body of the mill into the area of higher pressure there. When a sail passes into this sector, the cloth may be lifted off the supporting framework, causing it to flap. Not only does this weaken and damage the cloth but it causes any rain on the sail to be flung against the body of the mill.

Smeaton found that the velocity of the tip of the sail was nearly in direct proportion to the velocity of the wind. Therefore, while increasing the area of sail at the extremities of the arms would give greater leverage and power, there was little benefit in extending the radius because the revolutions would have to fall to maintain the tip speed ratio. He found that the best sails were rectangular with the surface on the driving side. He recommended extending the leading side with a thin triangular section which added about 25 per cent to the area, but then the angle of weather had to be increased. If the sail area, the 'solidity', were increased too much, 'so that then the whole cylinder of wind is intercepted, it does not then produce the greatest effect for want of proper interstices to escape'.[23] He does not seem to have examined the solidity on mills with more than four sails.

The effect or output was governed by the following rules. First, the velocity of the sail varied as the velocity of the wind so that the peripheral speed of a sail was nearly proportional to wind velocity (V). Second, the load or maximum effect at a given time varied as the velocity of the wind multiplied by the velocity of the sail so it varied as the square of the velocity of the wind (V^2). The effect, that is the power developed, at maximum was nearly proportional to the cube of the velocity of the wind (V^3).[24] The force exerted by the wind was nearly proportional to the square of the radius of the sails. Smeaton's formula for the wind pressure, $P = 0.005V^2$, was slightly inaccurate, for modern work has shown that the coefficient should be just under 0.003.[25] Smeaton went on to compare the output of a windmill with other forms of power and found that one he built to his improved design with sails of radius 9.14 m (30 ft.) was equal to 18⅓ men or 3⅔ horses, taking five men as equal to one horse, compared with only ten men or two horses for the best Dutch performance.[26]

Smeaton's results were to stand the test of time – at least, for many years. Probably no one else carried out such exhaustive experiments until the work of T.O. Perry in the United States of America in 1882–3. On the Continent, Smeaton's results were accepted by Poncelet in 1845.[27] W.J.M. Rankine, an authority on steam engines writing in the late 1850s, modernised Smeaton's mathematics but accepted his basic principles, saying 'The reduction of the art of designing windmills to general principles is almost wholly due to an experimental investigation by Smeaton.'[28] When the possibility of introducing windmills into India in 1879 was being investigated, Smeaton's treatise was used as a basis for certain calculations.[29] Even as late as 1895, it could be stated that Smeaton's experiments 'are still the most reliable data extant'.[30] Yet not everyone agreed with Smeaton for when James Ferguson published his lectures on mechanics in 1773, he wrote

> Lastly, these ribs ought to decrease in length from the axis to the extremity giving the vane a curvilinear form; . . . All this is required to give the sails of a wind-mill their true form: and we see both the twist and the diminution of the ribs exemplified in the wings of birds.[31]

The concept of a taper towards the extremity for the blades of a windmill would not be taken up until after the First World War.

Winding the sails

As soon as a rotor yaws out of the main windstream, performance deteriorates because the blades rotate in a cross-flow situation. In each revolution, the angle at which the wind hits the blade will vary all the time, giving rise to unsteady flow over the blades, resulting in loss of power and in periodic dynamic variations.[32] It is very difficult to calculate this deterioration but it is evident that a substantial amount of power can be lost through the wind changing direction slightly. Then greater stresses may be imposed upon the blades themselves in addition to the normal reversal of forces which occur through gravity as the blades rotate in each revolution. The incidence of such forces can be decreased both by keeping the mill facing the wind and also by spilling gusts from the sails in one way or another.

To what extent the problem of winding the mill was exercising minds of inventors in the first half of the eighteenth century is hard to discover but it may be significant that two patents connected with horizontal windmills specifically mention that such mills avoid this difficulty. In 1724, John Brent, who through John Calley was associated with Thomas Newcomen, described how his mill was 'so disposed to work with the wind blowing from any point of the compass, without turning or altering the position of the said engine or house thereto belonging'.[33] John Kay, better known for his flying shuttle on the loom, also pointed out how, on his mill, the sails 'receive the wind without shifting the engine, let it blow from any point whatsoever'.[34]

Edmund Lee, of Brock Mill near Wigan, provided the solution which would be applied to windmills and even wind turbines up to the present day (Figure 42). Part of his patent granted on 9 December 1745 contained the specification for 'a self-regulating wind machine' with 'back sails' known as the 'fan tail'.[35] The English patent contains no description but has a drawing showing a framework to the rear of a tower mill on which the fan or 'fly' is mounted. The fly drives a vertical shaft to gearing at ground level. Further information is contained in the Dutch patent granted by the States of Holland and West Frisland on 19 January 1747 (8 January 1746/7 in England).

> When this Self Regulating Machine is required to keep the Sails always in the Eye of ye Wind, it is Performed by 4 or 6 Smaller eliptical Sails fixed in a frame at ye opposite Side of the Mill to the Main Sails, in a Position cross [at right angles to] the Main Sails, so that when the Wind Shifts from ye Eye of the Main Sails, in the Smallest Degree, it Immediately Catcheth the Back Sails, the Turning of which, by a great Power [by a system of reduction] it Communicates to a travelling Wheel which moves round the Mill, conveys the Main Sails again into the Eye of ye Wind, Part of that Machine in which Said Back Sails are fixed, being fastened to that outward Part of the Mill which turns round.[36]

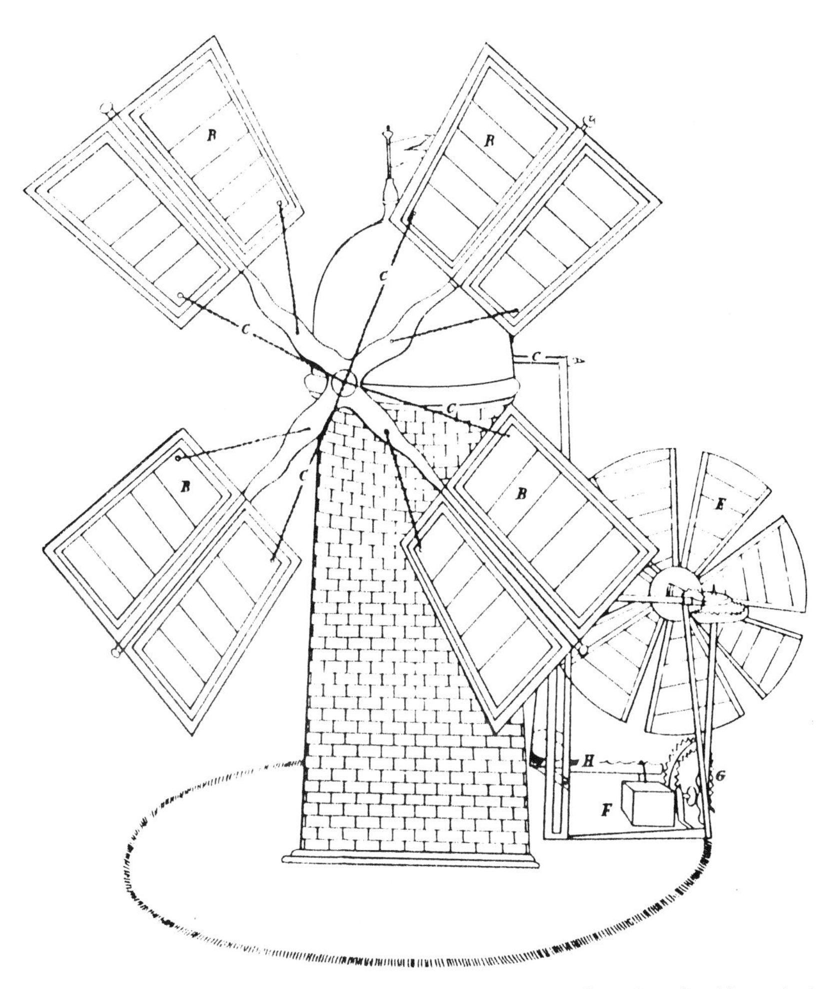

Figure 42. Edmund Lee's patent drawing of a windmill with fantail and self-regulating sails (1745).

The fantail blades on early examples were set at an angle of 40 to 45° to the shaft on which it rotates but later this angle was about 55°. When the wind blows at right angles to the shaft, nothing happens, but as soon as it changes to one side or the other, it plays on one face of the blades and turns the fly round.

Two thousand revolutions of the fly might be necessary to rotate the cap of a tower mill in one complete circle,[37] so that gearing was fitted to link the shaft of the fly. It has been suggested that the widespread introduction of the fantail had to wait for cast iron gearing and it may be noted that Brock Mill was a forge where Lee could have had parts made.[38] On post mills, something like the arrangement shown in Lee's patent was constructed. It was found unsatisfactory to mount the fan on the end of the tail pole as it was too far out and tended to act as a weather-vane and swing the mill from side to side. At the end of the ladder was built a carriage on which wheels, driven by the fly, moved round a circular track. In the case of post mills in Essex (Figure 43) or Suffolk, the fly was normally mounted on a vertical extension of the carriage so the drive shaft was quite short. As these driving wheels had to be set radially to the mill, some very curious forms of gearing resulted. In East Sussex, the fantail was sometimes mounted on the top of the buck, but then the drive had to be taken down the whole length of the buck, along the ladder and to the carriage. Tower mills dispensed entirely with tail poles so that the fan was mounted on a special framing at the rear of the cap. In this case the curb would have a gear ring formed around its top, which could be internal, around the upper surface or external. The external type was found most frequently with worm gears which were popular in Norfolk, Kent and Sussex.[39] Otherwise ordinary pinion gearing was used.

John Smeaton knew about fantails by 1782, for that January he wrote from his home at Austhorpe near Leeds to Josiah Wedgwood,

Figure 43. The post mill at Aythorpe Roding, Essex, fitted with a fantail on framing over the ladder with the tail pole shortened (May 1990).

> I must also advise you, that in this part of the country, it is a common thing, to putt Sail Vanes, that keep the mill constantly to wind, without attention or trouble to the millman.[40]

He fitted a fantail on Chimney Mill, Newcastle upon Tyne, which he built in the same year, and provided a crank on the drive shaft so it could be turned by hand in a calm.[41] In 1807, it was said that in the Humber area, fantails were generally applied to regulate the position of windmill sails and keep them facing the wind.[42]

However, the fantail remained essentially a phenomenon of the eastern part of England. While there is one on Wilton Mill, near Hungerford, Berkshire, few if any are known on the western side of the country except in the Fylde district of Lancashire. Possibly they were unsuitable for stronger winds or where wind direction might change suddenly. Rex Wailes pointed out that many a mill was tail-winded in a thunderstorm, which advances across the country like a wave curling back on itself when the hot air in front of it flows in from the opposite direction as it rises to meet the storm. At any point just in front of the line of advance, the wind, which will have been blowing in the opposite direction to the storm, drops and then will suddenly blow strongly with the storm as it sweeps on. This gives the fantail no time to turn the mill and disaster could follow. There was a similar problem in a storm when the miller might want to turn the sails out of the wind in order to stop the mill, for the fantail (unless it were disconnected) would be trying to keep the sails facing the wind. Some mills in the north of Holland and Germany as well as into Denmark were equipped with fantails but they were never popular elsewhere until the introduction of the American wind engine. A very few mills, such as Happisburgh post mill, Norfolk, were fitted with two, and two have been seen on a modern wind turbine in the north of Holland.

Regulating the stones

Desaguliers published the following comment in 1744.

> The Reader may perhaps wonder that in this Account of such Variety of Engines, I have given no Account of Wind-mills; but I would not do it, because the principal Thing is wanting in them; that is a Method to make them grind Corn uniformly, when the Wind suddenly varies: for sometimes from scarce bruising the Corn, the Motion is so increas'd, and the Stones go so fast, that the flour is quite hot and spoil'd. Diminishing the Surface of the Sails is practised, but that can't always be done quick enough for the sudden Increase of the Wind. There might be some Contrivance to give the Mill more Work to do, which it should take of itself, as the Wind rose suddenly, and leave as the Wind grew slack. I don't

> hear that any Body yet has made use of any such Contrivance. I am confident it may be done, and I have design'd it above these twenty Years; but have been hinder'd by other Affairs. But I hope to do it some time or other, if I live.[43]

Regulating the speed of windmills must have been concerning other people at the same time, for Edmund Lee claimed in his patent that ordinary windmills were

> Subject to great irregularities in their Motion & thereby not only frequently damaging the Mills & Engines, & the several Movemts, thereof, but also very dangerous & too often fatal to the Persons attending them, by reason of the sudden, high & unexpected Gusts & Gales of Wind.[44]

We do not know what quality of flour was acceptable and whether that from a windmill was poorer than that from a watermill with its better speed regulation. The windmill faced two problems: lulls in the wind, when the mill might not be able to grind; and too much wind, so it went too fast and, as Desaguliers says, might burn the flour. The first problem was partially solved by the centrifugal governor and the second by 'patent' sails regulated by a counterweight.

The miller had to engage the drives through the stone nuts before starting his mill, so he would judge the strength of the wind and engage as many sets of stones as he thought could be driven by the spread of sails. If he expected a gentle breeze, he would set full sail but put into mesh only one set of stones. In this way he could try to equate the power input with the power output. But the mill had to be started, and this would need a stronger gust, particularly if the runner stones had been lowered in the grinding position. So he would start with the runner stones up and lower them through the tentering mechanism when the mill was turning. When the wind dropped, the stones had to be lifted to reduce the load and keep the mill running to make the most advantage of the next gust when the stones had to be lowered again. While the damsel regulated the flow of grain into the stones according to the speed of the mill, some device was needed which would raise the stones at low speeds.

The first known attempt to regulate the gap between the stones was patented in 1785 by Robert Hilton, flour merchant of Preston, Lancashire.[45] A centrifugal fan in a windbox sent along a trunk or pipe a draught of air which varied according to the speed of the mill. It acted on a flap or baffle which was connected through ropes and levers to one end of the bridge tree carrying the bottom bearing for the drive to the stones (Figure 44). Movement of the baffle raised or lowered the bridge tree so that, as the speed increased, the upper stone could be brought lower to maintain a constant quality of flour. This device was soon replaced by the more efficient centrifugal governor.

In 1790, Robertson Buchanan sketched two 'lift-tenters' which he had seen operating in the neighbourhood of Liverpool, but gave no date for their inven-

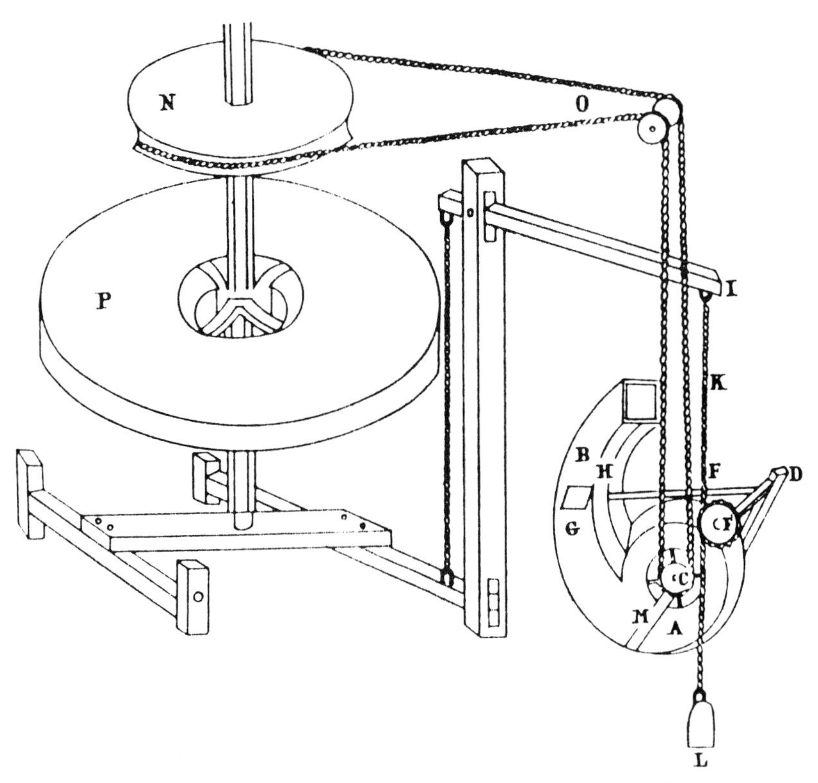

Figure 44. Hilton's speed sensing device with a centrifugal fan at C sending air along the tube GB where a flap worked the tentering mechanism (1785).

tion. He too commented on the quality of flour ground by a windmill and said that sometimes the grain was forced rapidly through without being sufficiently ground. The devices he saw used the principle of the centrifugal pendulum, the end of which moves out and up through the centrifugal force acting on it. The extent of this movement is proportional to the velocity. Levers from it can be linked to the bridge tree in a way similar to Hilton's baffle to control the gap between the milling stones. One of Buchanan's 'governors' had a single weight or pendulum and the other the more usual form of a pair.[46]

It is not known whether these governors preceded Thomas Mead's patent in 1787.[47] He used the 'lag' form of centrifugal governor mounted above the stones just below the driving nut. As the weights rose up through the increase in speed, so, through linkages, the bridge tree was lowered (Figure 45). He claimed it was 'a regulator on a new principle' and, many years later in 1825, Dr Alderson, stated that 'his late friend Mead, who, long before Mr Watt had adopted the plan to the steam engine, had regulated the mill sails in this neighbourhood [Hull] upon that precise principle'.[48] John Rennie, a millwright, fitted centrifugal governors to the stones at the Albion Mill in London, for Matthew Boulton wrote to James Watt in May 1788 about

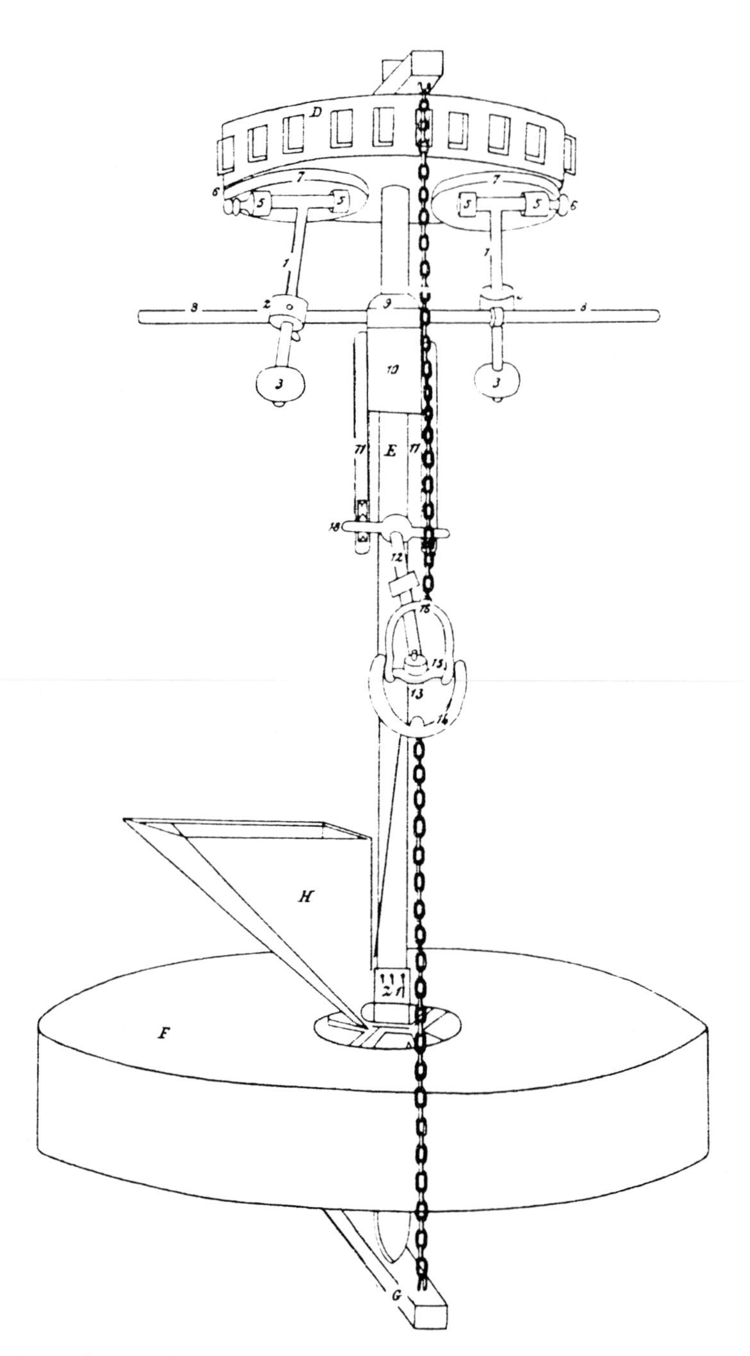

Figure 45. Mead's lift tenter with a lag type of centrifugal governor (1787).

> regulating the pressure or distance of the top mill stone from the Bed stone in such a manner that the faster the engine goes the lower or closer it grinds & when the engine stops the top stone rises up & I think the principal advantage of this invention is in making it easy to set the engine to work because the top stone cannot press upon the lower until the mill is in full motion.[49]

That December, Boulton and Watt adapted the centrifugal governor to control the admission of steam on the LAP engine which drove their Birmingham works, the first of very many such applicatons. Boulton had realised the advantages of the centrifugal governor for corn milling but may not have recognised just how appropriate was the movement that it gave. When the weights begin to move outwards from their lowest position with their rods almost upright, they give the greatest vertical movement to the linkage and so would lower the stones quickly at the slower speeds. At faster speeds when the arms to the weights are approaching the horizontal, there is less vertical movement and so the stones would have been closed more slowly. A variation of the centrifugal governor was applied to tentering in windmills in 1789 by Stephen Hooper (Figure 46). A novelty in this patent was an additional linkage to the hopper above the stones for controlling the flow of grain according to the speed.[50] These were the forerunners of a wide variety of governors which appeared in windmills during the nineteenth century.

Speed control through the sails

In strong winds, the amount of power developed could be regulated by twisting the blades so that they presented a different angle to the wind, by varying the area of cloth or by spilling the wind. Some control system was necessary which could be triggered either by the pressure of the wind or the speed of rotation. The earliest patent for some form of automatic regulation was taken out by William Perkins in 1744. Part concerned a machine for grinding corn, raising water and so on with an eight-sailed horizontal windmill. The sail was fitted with a balance weight which

> causes a sufficient resistance in the wind to do the work required, and no more; then if the wind blows harder than is necessary, the sail will turn and raise the lever with the weight a little, and let the wind through; so consequently a strong wind has no more power (as to the circular motion of the sails) than a wind that is just sufficient to do the work.[51]

This is a good description of the principle and operation of a weighted lever system for using wind pressure to stop a mill from overspeeding, but there is no drawing. In one form or another, this would become the most common and successful method for providing an upper limit to the amount of power being generated and one which is still used today. With low wind speeds, no control was necessary but the weight could come into play when pressure on the sails related to a certain wind speed was reached. This would prevent the mill overspeeding and relieved the miller of the fear of being caught with his mill spread with too much cloth. In 1787, Benjamin Heame included a fusee scroll in the linkage to the weight so that increasing force of the wind would be countered by the fusee shape.[52]

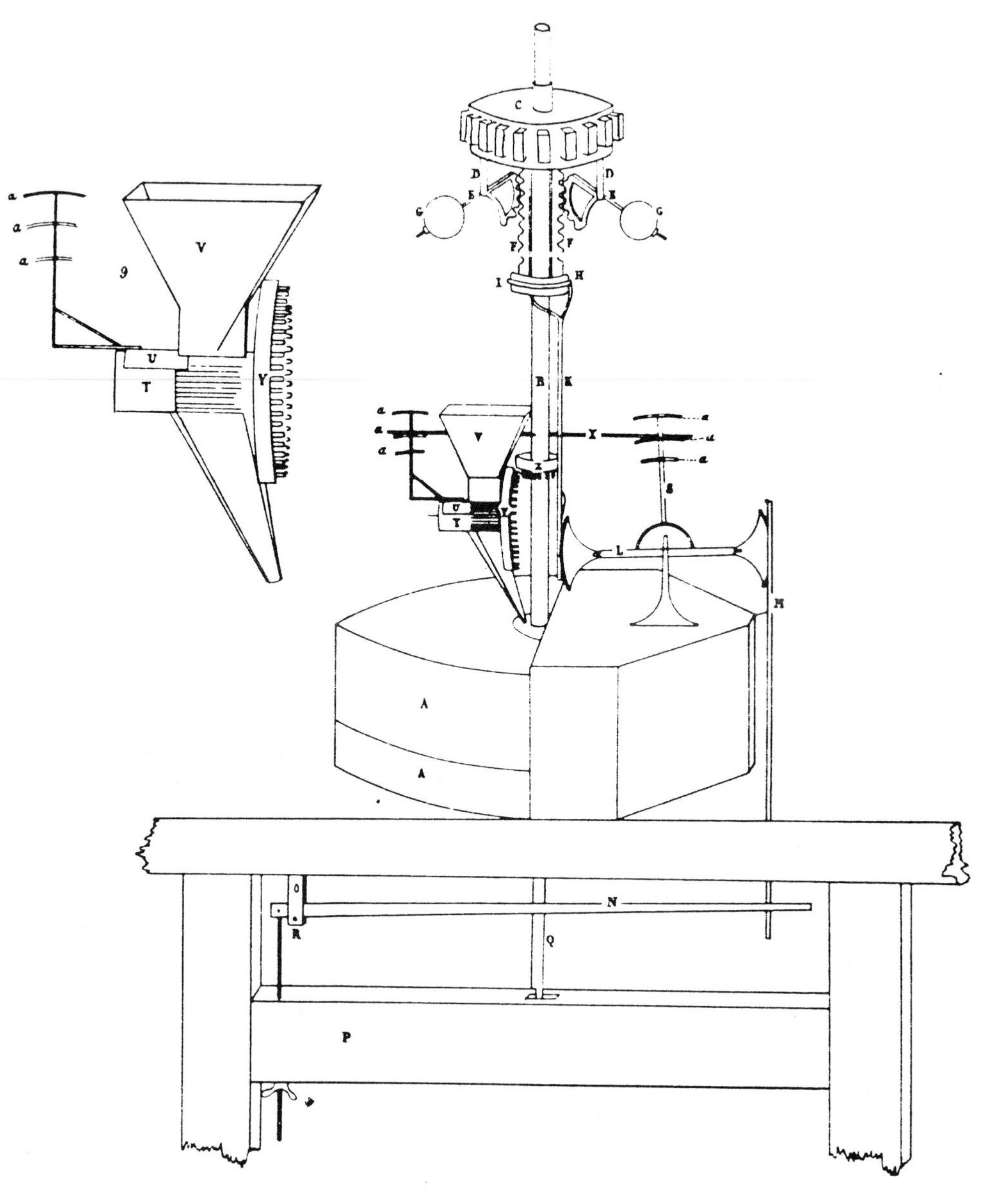

Figure 46. Hooper used the more conventional form of centrifugal governor with the weights moving outwards. There was one link to the tentering beams and another to a device to regulate the flow of grain to the stones (1789).

Although Edmund Lee is best remembered for his invention of the fantail, his patent envisaged a complete form of 'self-regulating wind machine' which not only turned itself so the sails faced the wind but was furnished with a 'weight which regulates the sails according to the wind force'.[53] His drawing shows a tower mill on which the weight controlled the sails by twisting them to alter the angle of weather they presented. While his method proved impracticable, the principle was important for later developments. His blades had equal leading and driving sides and at their bottom corners were fixed chains, the other ends of which were attached to 'a regulating barr passing through the centre of the original axes'. It was this 'striking rod' through the centre of the windshaft which would be copied by most other inventors of other self-regulating sails. The weight hung from a 'crooked Lever' at the other end of the striking rod and was proportioned

> to the Greater or less Force Required to perform the particular Work of Each Mill . . . This Weight being once justly proportioned there will, Afterwards, be No Necessity for any Alteration, let the Wind blow how it will.[54]

Lee claimed that the additional expense of erecting his mill would be very little above the common ones and 'the Expense of attending it & keeping it in Repair, much less'.

John Barber's sails, patented in 1773, were not successful either. His mill for operating a furnace for refining metals was shown with small oval sails in two circles, the outer probably with thirty-two and the inner sixteen, which anticipated the annular sail. He said they were

> never folded up as in the common windmill, but are kept at full breadth and hung upon centres so that they may turn with either a side or edge to the wind and are held to the proper weatherings by balance cords, which running down the arms of the wheel, are united at the main shaft and then passing through a hole in the centre thereof, pass over a pulley within the tower and suspend a box of weights, which being increased or decreased hold the vanes harder or slacker to the wind to the power required.[55]

In 1800, Robert Sutton tried twisting wooden blades worked by ropes and levers which could be turned to the 'position of greatest angle of weather for producing the best possible effect'.[56] On traditional windmills, this form of speed regulation was not a success.

Because they were another dead end in development for the traditional windmill, those inventions which varied the amount of cloth spread will be considered next. These patents occur at about the time when James Watt was perfecting his rotative steam engine. Boulton had been pressing Watt since 1781

> The people in London, Manchester and Birmingham are *steam mill mad*. I don't mean to hurry you but I think . . . we should determine to take out a patent for certain methods of producing rotative motion from . . . the fire engine.[57]

The increasing demand for power from the industries such as textiles during the Industrial Revolution created a crisis because the best waterpower sites were already occupied and water resources were scarce. If the windmill could be improved, it might help to alleviate the power shortage and erection of windmills in grain-producing areas may have freed waterpower sites elsewhere for other industries. For example, in 1784, Samuel Greg took over the water rights of a corn mill and established Quarry Bank Mill, Styal, Wilmslow, Cheshire, for spinning cotton.

The first of this series of patents is the first in which the principle of the centrifugal governor has been discovered. The fourth drawing in Benjamin Wiseman's patent of 1783 shows

> the machine by which the sails are let down when their velocity exceeds a certain rate. The principle of this machine is that tendency which all bodies moving in circles have to fly off from their centres of motion which is called centrifugal force, the increase of which force is as the square of the velocity.[58]

Underneath the great spur wheel, Wiseman suspended a weight on the end of a rod. It was connected to the vertical arm of a right-angle lever which was pivoted at the right-angle. A regulating weight could be moved along the horizontal arm and be set so it remained stationary when its action was 'just sufficient to keep it [the weight on the pendulum rod] from flying off when the sails rotate twenty times in a minute'. At any faster speed, the pendulum weight overcame the regulating weight so that the right-angle lever lifted, engaged with a trigger, which released the sails. This was more of an overspeed safety device than a true governor.

As well as his method of controlling the gap between the stones by an air regulator in his 1785 patent, Robert Hilton also addressed himself to the input side of the problem by patenting a method of furling the sails. Along the leading edge, he placed a roller which could be turned to wind on the sail cloth. The trailing edge had ropes connected to a second roller which could be turned to pull out the cloth. Other ropes passed round the furling or unfurling 'purchase wheels' which could be turned through gearwheels connected to a common gearwheel on the end of the windshaft. This gearwheel was brought into play by the miller through a friction drive operated by rods along the windshaft.[59] It was a very complex mechanism and two years later, Thomas Mead, a carpenter from Sandwich, Kent, tried to improve it by fitting a true centrifugal governor (Figure 47).

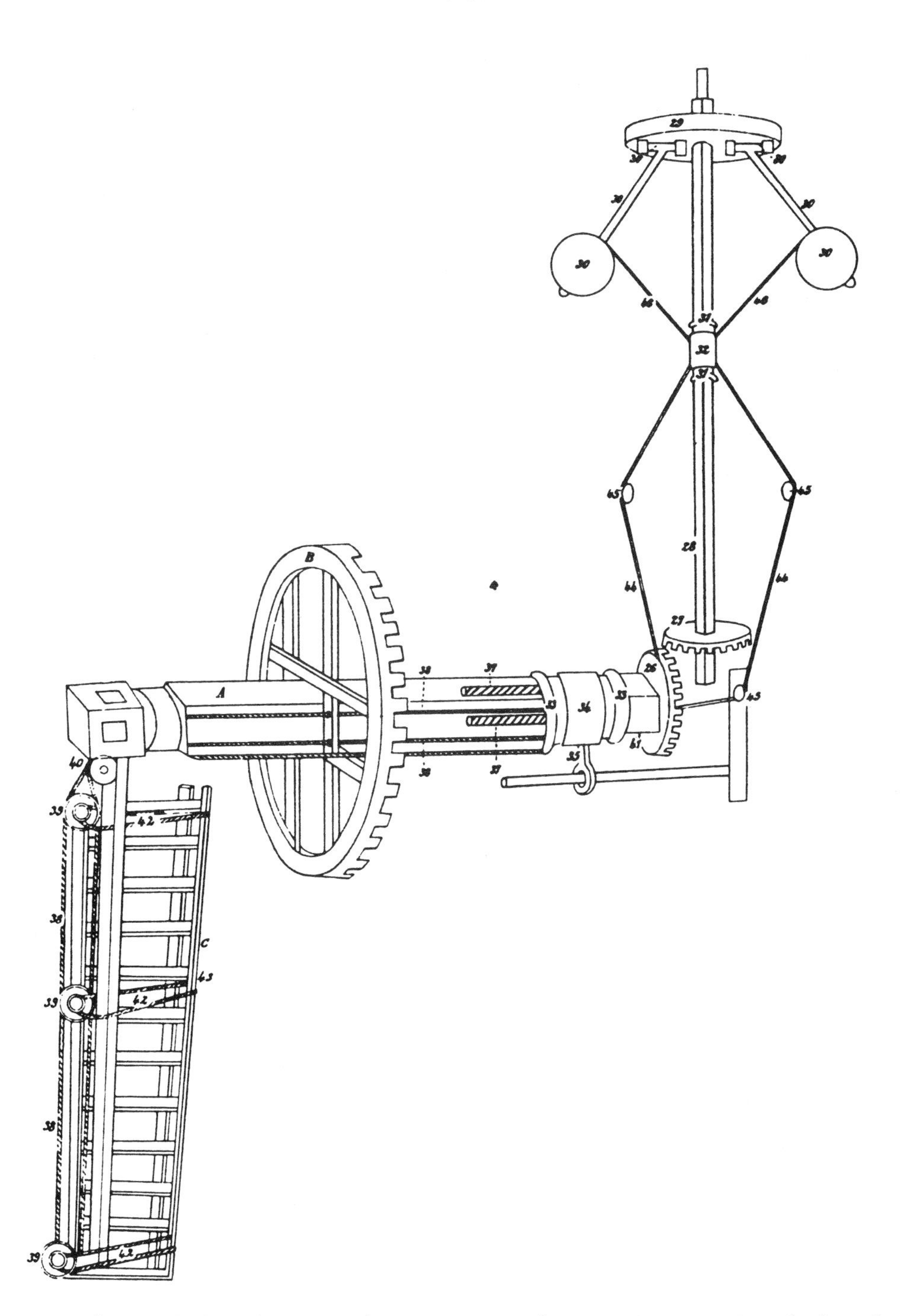

Figure 47. Mead adapted the centrifugal governor to draw the cloth over the sails through a complicated system of ropes and pulleys (1787).

> As the velocity of the mill increases, the globes fly off from the centre of the bar and draw up the collar on the bar, thereby drawing the collar on the shaft nearer the pulley at the end, and also drawing one end of the cord which turns the furling pullies, and the sails are thereby furled up in proportion to velocity. For unfurling the sails, there are springs which pull the collar on the shaft nearer to the sails.[60]

The various cords or ropes would have been difficult to keep tight and furling or unfurling the sail across the framework of the blade while the wind was blowing must have been very difficult too. It is not surprising to find that Hilton's complex system, even though improved by Mead, was little used. It is more surprising that the centrifugal governor was rarely applied to the speed control of a windmill through the sails even after better apparatus had been invented. Perhaps early governors were not sensitive enough and had insufficient power to operate the mechanisms.

An improved form of roller furling mechanism was introduced by John Bywater of Nottingham in 1804. Instead of ropes, he used gearing which would have given a more positive action, but his mechanism seems to have been put in and out of gear by the miller pulling on ropes. The cloth could be folded like a curtain instead of rolled but Bywater preferred to roll up the sail. Also a centrifugal governor could be fitted to regulate the mill should the miller be absent.[61] Bywater formed a partnership with Thomas Simpson and certainly Carlton Mill, Nottinghamshire, was fitted with their sails. This partnership was dissolved in August 1807.[62] A similar sort of mechanism is shown in the windmill drawing published in 1806 by Andrew Gray in his book on millwrighting.[63]

More successful were Stephen Hooper's roller reefing sails patented in 1789[64] with a series of small roller blinds, each one across the width of the blade (Figure 48). The cloth on each roller was connected by two webbing straps or 'listings' to the next roller so that, as it wound onto its own roller, the listings were unwound from the next and vice versa. All the rollers were operated at once by rodding along the length of the whips which in turn was connected through the windshaft to the interior of the mill where the system was controlled by weights and counterweights to make the regulation automatic. Some mills were fitted with roller reefing sails, one of them being Hessle Cliff Whiting Mill near Hull, which operated with them until 1925.[65] One weakness of the apparatus lay in the listings connecting the blinds.

The method which became the most popular used a form of Venetian blind which could be turned to allow the wind to spill or pass through. This originated with Andrew Meikle, who sent John Smeaton a drawing on 17 March 1772 of an 'Improvement proposed on the sails of a windmill so as to answer the inequalities of the power of the wind'.[66] The length of the blade was to be divided into sections and across each section was fitted a canvas-covered frame. The frame was pivoted at a point about a third of the way along the shorter

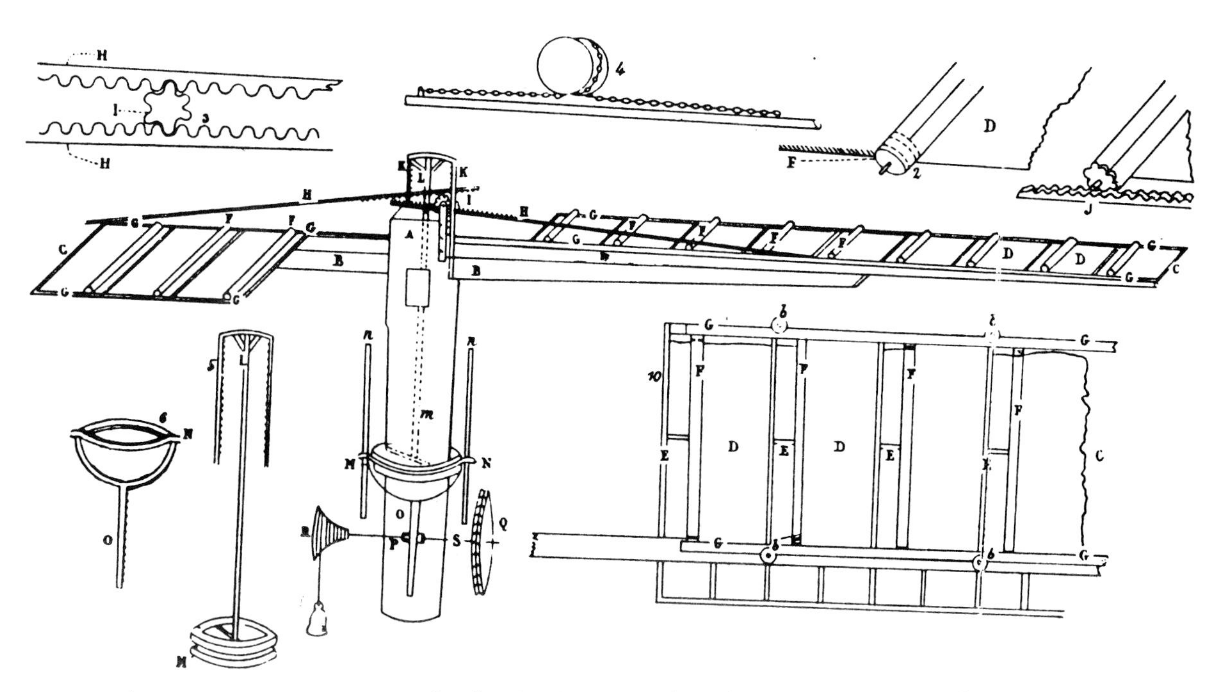

Figure 48. In another patent, Hooper fitted roller shutters on the sails in an attempt to regulate the speed by varying the amount of cloth exposed (1789).

sides. At the end by the whip, each frame had a wheel on this pivot which was connected to its own spring. When the strength of the wind increased, it would overcome the force of the spring and open the shutter which allowed the wind to pass through. As Meikle said, 'When the wind blows very hard, it will so counter-balance the spring that the frame will turn its edge to the wind and so the mill will be drove about with less violence.' When the wind fell, the shutter would be closed by the spring. When actually fitted to windmills, the 'spring sail' had all the shutters connected to a rod or 'sail bar' which was linked to a single spring for each blade. The tension in this spring could be set by the miller at the beginning of the day's work to suit the power he needed and the strength of the wind. The pivot was moved to one end of the shutter, possibly on Smeaton's advice, so that the whole of the shutter opened away from the wind.

Meikle's spring sails can be found on surviving mills in many places. The springs vary from coil at Wrawby, Lincolnshire, elliptical at Stocks Hill Mill, Wittersham, Kent, or quarter-elliptical at the White Mill, Sandwich, Kent (Figure 49). While they gave an overspeed control and protected the mill from sudden gusts, each spring had to be tensioned individually which still meant that every sail had to be brought into the lowest position for setting when starting or stopping. Once set, there was no way of making adjustments when running although they were practically self-governing. When spring sails passed into the high pressure zone at the front of the mill, they would shut with a snap, which, in wet weather caused a shower of water to spray over the mill. Quite often mills were fitted with two common cloth sails and two spring sails because the shuttered sail never gave the power of a well designed common sail. This combination gave a fair degree of regulation together with fair driving power, particularly in lighter breezes.[67]

The final solution of an overspeed regulator for the traditional windmill was provided by a Norfolk man, William Cubitt, in 1807.[68] He patented an improved form of Meikle's sail so that it was not only self-regulating when running but could be easily controlled by the miller whenever necessary without stopping the mill. Meikle's shutters were connected to sail bars which in turn were joined to a 'spider' at the end of the windshaft by links and bell cranks (Figure 50). From the spider through the windshaft ran a control or 'striking rod' which terminated in a rack and pinion. To start the mill, the miller pulled on the appropriate side of a continuous rope or chain round a wheel on the axle of the pinion which moved the rod and closed the shutters. He had selected an

Figure 49. On the White Mill, Sandwich, Kent, the spring shutters are controlled by quarter eliptic springs near the windshaft which are tensioned at the tips of the sails (April 1991).

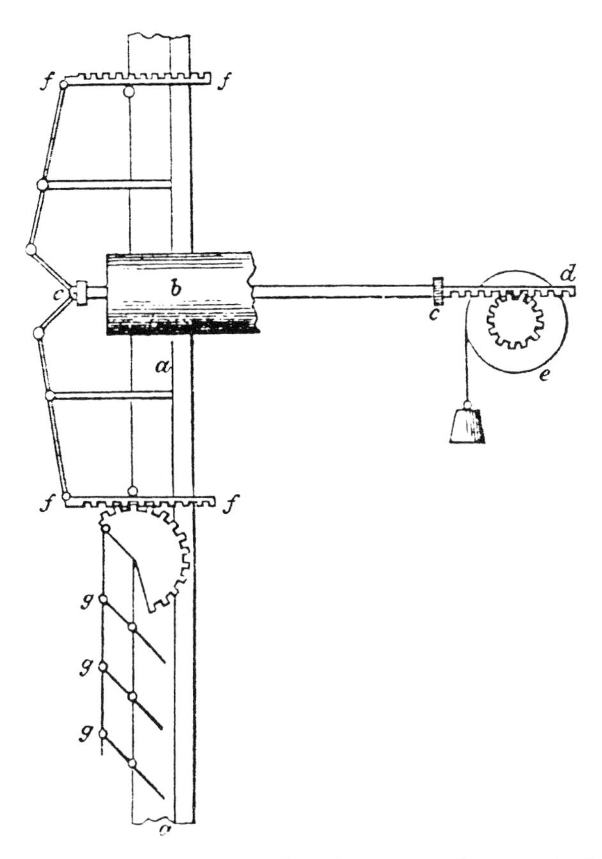

Figure 50. The weight operating through a rack and pinion has been retained subsequently on Cubitt's patent sails but on the sails bell cranks have replaced the racks (1807).

appropriate weight for the power he wanted which he hung on that side of the chain so that, in a gust, the force of the wind on the shutters would overcome the weight, raise it through the rack and pinion and so spill the wind by opening the shutters. When that gust had passed, the weight would fall and close the shutters again. To stop the mill, the miller merely had to remove the weight from the chain and open the shutters. Cubitt had been partially anticipated in this way of using weights by Robert Sutton in 1800. Instead of a rack and pinion, Sutton, a clockmaker from Barton upon Humber in Lincolnshire, used a linkage with levers controlling the shutters in fan-shaped sails. He also suggested roller bearings on the windshaft 'which, by annihilating all friction, is an incredible assistance to the motion of the sails'.[69]

The first mill fitted with Cubitt's 'patent' sails is said to have been Horning post mill in Norfolk where the new sails were supplied by the Ludham millwrights, Englands.[70] Another contender was a smock mill at Stalham, Norfolk owned by Cubitt's father-in-law, which had sails 8.53 m (28 ft.) long by 3.99 m (13 ft. 10 in.) wide, with four rows of twenty-seven shutters or vanes with canvas-covered wooden frames, making a total of 432 shutters. In this case they were controlled by friction driven governors from the windshaft.[71] Cubitt licensed millwrights in many areas to supply these sails. When patent sails were fitted together with Lee's fantail, the mill became virtually self-regulating as far as the wind was concerned. This allowed the miller to concentrate on production and helped him compete with the new steam-driven roller mills towards the end of the nineteenth century (Figure 51).

Another development to traditional sails was made by a Suffolk man, Robert Catchpole, from Sudbury, a little before 1860. At the outer ends of leading edges of a patent sail, he fitted longitudinal shutters which, when closed, gave additional driving power at the most effective point. They were controlled by the normal shutter mechanism so that, when the shutters opened, these 'sky-scraper' shutters also opened and acted as air brakes, spoiling the flow of wind round the sails. The post mills at Combs Gedding and Wetherden and the tower mill at Sibton were fitted with them.[72]

Traditional multi-sailed mills

In western Europe, there were some mills with more than four sails and one with six is recorded at Framlingham in 1279.[73] Leonardo da Vinci sketched a tower mill with a rotating cap and the stocks for six sails but he did not include the blades so we do not know whether it would have had the framed type of northern Europe or the triangular cloth type of the Mediterranean.[74] In 1726, there was a six-sailed windmill for supplying water to London.[75] However, making holes for a number of stocks seriously weakens the end of a wooden windshaft. The multi-sailed mill with traditional northern European blades had to wait for the introduction of the cast iron windshaft. John Smeaton was the first person to make axles for a waterwheel from cast iron and began using the same material in windmills when he erected the smock mill at Wakefield in 1755.[76] Cast iron would transform the gearing in mills and also enabled stronger windshafts to be manufactured. In East Anglia, massive windshafts were cast with square, box-shaped canisters on their ends through which the stocks were passed, but these carried only four sails. For more sails, a cross was cast which was keyed onto the windshaft (Figure 52). The sail backs for each individual sail met in the middle and were bolted onto the face of the cross. This method reduced the need for massive pieces of timber to form the stocks. In 1774, Smeaton fitted five sails onto his flint mill in Leeds and also on the last two windmills he built at Austhorpe in 1781 and Newcastle upon Tyne in 1782.[77]

It has been claimed that a five-sailed mill is the smoothest because four sails are still driving the mill when the fifth passes in front of the body and loses the

Figure 51. North Leverton subscription mill is still operating with a full set of patent sails, fantail and ogee cap. The whips or stocks are bolted onto an iron cross on the end of the windshaft. The chain and weight for controlling the patent sails hang down to the left of the mill (August 1991).

wind (Figure 53). However, should anything happen to one of the sails, the mill is out of action until it can be repaired, because if one sail is removed the rotor is unbalanced. A six-sailed mill has the advantage that, if one sail is damaged, it can still be balanced to work under less power with four, three or even two sails (Figure 54). Six-sailed mills became quite popular after the middle of the nineteenth century and there were even a few with eight, of which Heckington alone remains now (Figure 55).

The greater the number of blades, the greater the initial starting torque, which is an advantage in light winds, but there is not that much advantage when running because the speed has to be reduced. One blade must not interfere

Figure 52. The iron cross, on the end of the windshaft, shows how the stocks for six sails can be bolted on it to meet in the centre (see also Figure 53). The stone tower of Wymondham Mill, Leicestershire, may have been raised with a brick section when these more powerful sails were fitted (July 1991).

Figure 53. Maud Foster Mill, Boston, Lincolnshire. The five sail backs are bolted onto the face of the cross. Since this photograph was taken in April 1977, the mill has been restored to working condition and grinds corn once more.

with the air-flow of others which happens as the speed increases. The speed of a windmill varies inversely with the square root of the number of blades, so a four-bladed machine should run at about 71 per cent of the speed of a two-bladed, a six-bladed at 58 per cent and an eight-bladed at 50 per cent.[78] The sail surface on a traditional windmill with four sails usually occupies about 25 per cent of the swept area and the speed of the tips when working at maximum efficiency is about 2.5 times that of the wind. When the relative area is increased with more blades, the circumferential speed has to be reduced so that a greater number than four blades involves unnecessary expense and complication.[79] This is why multi-sailed mills never became very popular.

Figure 54. Trader Mill, Sibsey, Lincolnshire, is one of the few left with six sails. The fine ironwork round the staging should be noted (April 1977).

Figure 55. Heckington Mill, Lincolnshire, is the last of the rare eight-sailed mills (March 1990).

Annular sails

The annular sail is generally associated with American windpumps, but the concept may be traced to Mediterranean windmills as well as some English inventions. John Barber's 1773 patent[80] has been mentioned earlier. In 1832, John Burlingham described his windmill where the space swept by the sails was 'appropriated to the reception of any convenient number of vanes or sails so that they form nearly one continuous circle'.[81] In 1838, Henry Chopping, an Essex miller, set up a small model with an annular sail at his father's mill at Matching which he demonstrated to several local millwrights.[82] One of them told Richard Ruffle, a local miller, who built a mill and fitted it with an annular sail. After the collapse of this mill, in 1855 Ruffle built a more substantial one at Haverhill, Suffolk, on which he fitted an annular sail 14.6 m (48 ft.) in diameter with 120 shutters. Each shutter tapered from 31.5 cm to 35.7 cm (12½ to 14½ in.) and was 1.52 m (60 in.) long, made of a wood frame covered with canvas. This mill was demolished in the Second World War because it interferred with the flight path of a nearby aerodrome.

Henry Chopping fitted his own mill at Roxwell, Essex, with an annular sail 15.8 m (52 ft.) in diameter after it had lost its earlier sails in a gale. He estimated that it generated some 20 h.p. A group of Americans who visited him suggested that he should patent his design, which he did in 1868 under the names of Frederick Warner and himself.[83] The ring of shutters was supported by eight radial arms (stocks), and along every alternate one passed the shutter control rod which was operated through a spider and striking rod in the usual way. These radial rods were connected at their outer ends to others which formed chords linking all the shutters in their section. There was an additional spring-loaded device which enabled any section to open and spill the wind if a strong gust should hit that particular area. The patentees claimed that this would prevent injury to the sails and the shutters were designed to blow out should the mill become tail-winded. Chopping assigned his patent to John Warner & Sons Ltd, who added windmills for both pumping and grinding to their catalogue. A mill for the Newhaven and Seaford Water Company at East Blatchington, Sussex, in 1892 was a failure because it was badly sited and salt water contaminated the well.[84] But one of their mills was illustrated in the 1910 edition of the *Encyclopaedia Britannica* and shows the links between English and American practice.[85]

CHAPTER 6

Windmills for land drainage

The Netherlands

Drainage machinery

After propelling sailing ships and driving corn mills, the power of the wind was next used to turn windmills for draining water off the land. This happened first in the Netherlands. The greater part of the present Netherlands consists of a delta area formed from alluvial deposits brought down by rivers including the Maas, Rhine and Waal. Originally the Rhine reached the North Sea near Katwijk but around AD 700 the Lek branched off west of Utrecht towards Rotterdam and this became the main channel with the old course silting up in the twelfth century. The land, if such it may be called, was protected from the sea by sand-dunes along the coastline so that beds of peat grew up on it and these marshes rose above the level of the sea. However, the sea broke through the natural defences, inundating some areas, washing away the peat and forming lakes. When people began to inhabit this region, they built islands from the local clay on which they could live. It has been estimated that the hundreds of artificial mounds in the northern Dutch marshes were constructed from 100,000,000 cubic yards of clay: no small achievement with the labour resources available especially when compared with the 3,500,000 cubic yards of the pyramid of Cheops.[1]

Holland was mainly a dry peat area and cultivation for wheat was started around AD 1000 with the aid of drainage ditches. However, through flooding possibly caused by peat digging, and through shrinkage of the peat soil caused by drainage, conditions in the west and south of the Netherlands deteriorated. The first dam in a river for protection against high tides was erected in AD 802. Primitive systematic building of embankments and other artificial works surrounding the land originated sometime before the eleventh century although there are doubts whether this was undertaken primarily to protect existing areas from inundation or to reclaim land. After a severe flood in 1134, the first systematic collective defence against flooding was developed in the southwestern part of the Netherlands and it was here that the idea of polders originated. The first drainage sluice was mentioned in 1155. During the thirteenth and fourteenth centuries, people started to construct drainage canals and raise more banks. Comparatively primitive construction methods developed into regulated designs to suit different conditions, such as protection against the open sea, an estuary or a river. Much of this land, once embanked, drained naturally into a river or the sea at low ebb tide but, in some places, resort had to be made to artificial means.

Villages began action for collective protection against inundation and regional corporations came into existence so that gradually a complete administrative and judiciary system was evolved to control drainage and flood protection schemes.[2] These were important for the introduction of windmills which would have been beyond the resources of individual farmers. Districts were organised with a charter which determined the level of the water and specified which landowners had votes on drainage committees, who regulated voting rights, carried out inspections and supervised administration. Generally the flow of neighbouring rivers was too slow to be harnessed for working pumps operated by waterwheels. To begin with, men or animals provided the only power for raising water. Men used buckets, ladles or scoops to hurry the water along the ditches and throw it over the banks. Then there was the '*shaduff*' – the bucket on a rope lowered from the end of a counterbalanced pole – which was better known for raising water from a well. The ancient Egyptians had vertical wheels with pots around the circumference, driven by donkeys or oxen. The pots filled at the bottom of their travel and emptied as they turned over at the top, so lifting the water. Egypt was the home of the Archimedean screw and possibly lift pumps, too, which are the more likely types of early pumps in the Netherlands.

The drainage device which would assume the position of greatest importance in the Netherlands was the scoopwheel, which works like a waterwheel in reverse (Figure 56). The first definite mention of one does not occur until 1441,[3] but it is assumed that this was the water-raising device driven by the first windmills. While some forms of waterwheel might operate more or less in the open with water flowing over or under them, the scoopwheel must be housed within a structure forming sides and a breast curved to the radius to prevent the water running back into the lower channel. This structure had to be close-fitting to reduce leakage but it could support the bearings of the horizontal axle running through the middle of the wheel and also the gearing needed to drive it. The ladles, boards or scoops of the wheel had to be inclined to prevent the water, as it was being lifted, from running back over the tops. To prevent the water from the higher level falling back, a self-acting sluice door was placed in front

Figure 56. It is unusual to see a scoopwheel without a casing which prevents water splashing everywhere. This example is on the Hellouw wipmolen (March 1992).

of the wheel, which opened when the wheel was turning and raising water and automatically closed again when it was stationary.

The scoopwheel itself was similar in its construction to other parts of a windmill and could be kept in repair quite simply. It had the disadvantage that the distance through which it could raise water was limited to a fraction of its diameter. Walter Blith, writing in the 1640s, pointed out that the wheel 'must be made to that height as may be sure to take out the bottom of the water, and deliver it at the middle of the wheel'.[4] But, when allowance was made for floods in the river at the upper level and the dip of the ladle in the drain, the effective lift was much less than he implied. In 1710, it was reckoned that a 4.88 m (16 ft.) diameter wheel would give a lift of only 1.22 m (4 ft.) because, while the ladles could be immersed 1.52 m (5 ft.) deep in the drain at the start of pumping, they had to remain covered to 30 cm (1 ft.) when the drain had been reduced to its lowest level.[5] Owing to the problem of the water running back over the tops of the ladles, the effective lift was limited to about one third of the diameter of the wheel. Therefore the output to weight ratio was quite low.

There were other problems. One was that, while the slope of the boards helped the water to run off at the high level, they entered the water at the bottom too nearly horizontal and so did not dip in cleanly. Then there might be considerable churning of the water at the exit, particularly if the river were in flood, which wasted energy. Also, scoopwheels had to turn quite slowly to avoid centrifugal cavitation, but this closely matched the speed of traditional windmills. When in the 1830s Joseph Glynn was installing steam pumping engines in Britain, he found that the best peripheral speed for a scoopwheel was 1.83 m (6 ft.) per second because that helped to hold the water against the breast but did not fling it too high at the point of exit.[6] As the drain level dropped, there was less water to raise so the mill tended to turn more quickly. The water might not flow down to the scoops fast enough to fill them so the load would lessen further and, in a sudden stronger gust, the mill might run away. When starting with a full drain, there had to be sufficient wind to overcome both the inertia and the friction of the mechanism and also the dead weight of water already in the wheel. In 1926 when tests were being carried out on the Nieuwland Mill to the north of Schiedam, considerable difficulty was found in starting unless the wind was blowing strongly.[7]

The only major alteration to the scoopwheel was invented by Eckhardt. In his British patent, taken out in 1772, he claimed that his wheel would 'raise double the quantity of water and to double the heighth of any other wheel of the same dimensions then in use'.[8] He inclined the wheel at an angle of about 40 degrees and the ladles were specially shaped to fit into the race which also had to be built on a matching curving incline. The upper wheel bearing could be quite small while the lower one was contained in a vessel full of oil which reduced friction.[9] Eckhardt also tried to reduce the starting load by putting a door in the lower part of the wheel race which could be opened to allow the water to flow back into the drain.[10] Trials of his wheels were carried out in Holland between 1774 and 1776 when two mills of the old type and two mills of the new, all having the same dimensions, were constructed.[11] The ladles on the inclined wheels entered the water more smoothly and the inclination of the race enabled the water to be lifted as high as 3.66 m (12 ft.) without any increase in the diameter of the wheel. In the first experiments it was found that, with a strong wind, the inclined wheels evacuated a greater quantity of water than the common ones but in lighter winds they did not perform so well. It was claimed that five mills with inclined wheels gave a performance equal to twelve of the old but the type never became popular, either in the Netherlands or in England.[12] There is a picture of one of these wheels being used as a dredger.[13]

The Archimedean screw, driven by men or horses, may have been used quite early in the Netherlands.[14] It had the advantage that it could lift water higher than a scoopwheel but it was more difficult to construct. It had to be built up from small pieces of wood mortised vertically into the wooden axle shaft. If they did not fit closely, water would leak through the gaps. The upper working faces forming the screw were smoothed to assist the flow of water. Earlier screws were encased in a wooden cover which also had to be watertight to prevent leakage, a complex construction for the millwright. Between 1560 and 1700, as many as 102 patents for drainage mills were granted by the Dutch States General and individual states, besides numerous patents for other forms of pumps such as screw pumps, spiral pumps, centrifugal pumps and the like. In 1589, Cornelis Dircksz Muys patented a drainage mill driving a scoopwheel, which has been claimed mistakenly as an attempt to use an Archimedean screw.[15] Dominicus van Melckenbeke of Middleburg was granted a patent for a drainage device using a screw in 1598,[16] but the application of this type of water raising device in a special windmill, the '*tjasker*', can be traced back to Friesian manuscripts as early as 1580 in that region, where it was mainly used.

The tjasker was a primitive form of windmill with four sails mounted on one end of the windshaft just in front of the breast beam bearing which was carried on an 'A' frame (Figure 57). The shaft itself was extended to form the axis of an Archimedean screw enclosed in a casing. There were two layouts. In one, the lower end of the shaft rested on a bearing in the middle of a central round pond out of which the water was to be raised. The pond was fed by a channel passing under the higher drainage canal. This canal encircled the tjasker at the point where the Archimedean screw terminated and ejected the water. Surrounding this canal was a circular track round which the 'A' frame carrying the main bearing could be pushed so that the sails faced into the wind. The windshaft also carried a brakewheel.[17] The length of the stocks was 5.18 to 6.10 m (17 to 20 ft.), so it was quite a size and cumbrous to push round. In the second variety, the central pivot was situated at the top of the Archimedean screw in the middle of the pond at the higher level while the bottom of the screw dipped unsupported into the low level circular channel. The tjasker was never developed further.

A patent was granted in 1634 to Symon Hulsbos of Leiden for an Archimedean screw without a casing.[18] A brick trough reaching into the water to be lifted was constructed in which the screw revolved and a sluicegate was placed at the higher end to stop water flowing back. The screw fitted the sides of the trough closely and so presumably was simpler to construct than the enclosed type (Figure 58). These screws were angled at about 30 degrees with a diameter of 1.50 to 1.80 m (5 to 6 ft.). They raised the water 4 to 5 m (13 to 17 ft.), much higher than scoopwheels, and at 40 to 50 r.p.m. would lift about 40 cu. m per minute.[19] Their speed had to be considerably greater than that of the windshaft.

Polder mills were equipped with this type of Archimedean screw towards the middle of the seventeenth century.

Lift pumps were also tried. Again, their early history is obscure. In 1597, the famous Dutch millwright, Cornelis Cornelisz of Uitgeest, patented a drainage mill. His sketch shows a tower mill with a vertical shaft driving a horizontal one at ground level which was extended outside the mill and over a drain where it became either a double or a triple throw crankshaft for operating vertical pumps.[20] While not common on windmills, the first steam pumping engines could operate no other form of pump and, even when rotative engines were available, the three chosen for the drainage of the Haarlemmermeer in 1843 operated lift pumps. But lift pumps were better suited for lifting a little water to a great height rather than a great deal of water a small height, as was needed in land drainage.

Figure 57. A Tjasker windmill and Archimedean screw preserved at the Arnhem Open Air Museum, Netherlands (June 1972).

Figure 58. The construction of this 'open' Archimedean screw at the Schermer polder museum shows the many small pieces of wood (August 1990).

The introduction of windpower

Origins of the windpowered marsh mill remain obscure. A document from the reign of Count William IV (1344) mentions the sale of corn mills and wind-driven watermills in Drechterland,[21] and there was a drainage mill in Brielle in 1394,[22] but it can not be confirmed, through the terminology employed, that either of these were windmills. There is another possible reference in accounting records in the reign of William VI for the period between March 1406 and June 1408 when the dike-reeves of Delfland went to Alkmaar to inspect 'the water, which Florents van Alkemade and Jan Grietenzin throw out with the mill' and it states that 'Sir Floris will retain the wind of his watermill'.[23] This must have been a drainage windmill. Another reference concerns a drainage mill in the district of Zoeterwoude in 1408 and on 31 December 1413 yet another at Schipluiden (Delfland) where Philips de Blote, bailiff in Delfland, had a combined water and corn mill.[24] References to mills 'throwing water' occur more frequently as the fifteenth century progressed, particularly after 1450, but even in 1514 there were only eight marsh mills to the north of Amsterdam.[25]

Little is known about these mills, which must have been some form of tower mill probably built from stone. The problem was how to drive a scoopwheel fixed in one position while the sails had to turn to face into the wind which might come from any quarter. A post mill was unsuitable because all the machinery turned with the buck but, in a tower mill, the vertical main shaft could be extended to drive pumping machinery through a second set of gearing in the bottom. Pictures showing tower mills with conical caps, situated at the junctions of drainage channels and canals, survive from the sixteenth century so there is no doubt that such mills were used for drainage purposes. None of these early mills survives but their design must have been similar to that of the corn-grinding mill at Zeddam. Smock mills, most likely with open framing, are mentioned for the first time in 1438 although they originated probably before 1422.[26]

The Bonrepos mill in Noord Zevender near Schoonhoven is mentioned in a charter dated 13 May 1430, when Countess Jacoba van Beiren signed a document approving the construction of a drainage mill. From later evidence, it is assumed that this refers to a wipmolen.[27] However, the first definite mention of a wipmolen is in 1526.[28] In the wipmolen, the supporting post of a post mill was made hollow so that a vertical shaft could pass through it, connected by gears at the top to the windshaft and at the bottom to a further set which turned the axle of the scoopwheel (Figure 59). The structure supporting the hollow post needed to be larger and stronger than on the ordinary post mill and gave plenty of space inside for the gearing. At first, this lower structure was left open. Because the 'buck' of the wipmolen contained only the windshaft, the brakewheel, which doubled as a gearwheel, and the wallower or gearwheel at the top of the main shaft, it could be quite small.[29] Tail poles reached down to the ground for winding the sails and a ladder was fitted to enable the miller to climb into the top when necessary. The winding winch prevented the top rotating while the mill was working, which it might have done, and not the scoopwheel, through the torque on the gearing, something avoided with an ordinary post mill. The wipmolen was a much lighter form than either the tower or smock mill and so was more suitable for use in marshy areas where good foundations were lacking. Later the lower part of the structure supporting the post was enclosed and became a house for the miller. This type proved so efficient that by the end of the fifteenth century there were over seventeen large wipmolens pumping water into the region of the Vlist.[30] Soon the wipmolen was to become one of the characteristic and attractive sights in South Holland

Figure 59. This wipmolen, De Vrouw Venner, was rebuilt in 1913. It contains a house in the base and the tailpoles and winch can be seen on the right (March 1992).

but, although they were made larger and larger, they never attained the size of smock polder mills.

Larger windmills for drainage

A date of 1526 has been given for the introduction of the smock polder mill which spread from North Holland into South Holland.[31] In comparison with the wipmolen, the body of the mill became bigger and the cap smaller so that it was supported on a ring round the top of the body itself (Figure 60). Through greater size, not only was the living accommodation improved but, more important, larger sails could be fitted to develop more power. The earlier form still survives in North Holland, where the mill was winded by a winch inside the cap. In South Holland, the winch for the winding gear was attached to the bottom of tail poles (Figure 61). When a rope was connected to the brake lever in the cap, the miller could control his mill from the ground without having to climb it. Early polder mills look attractive with the reed thatch on their sides and caps and they have become the characteristic type of Dutch drainage mill. The rate of land reclamation was greatly increased through these new mills and a new era started around the middle of the sixteenth century. Between 1542 and 1548, the Dergmeer, the Kerkmeer, the Kromwater, the Weidgreb and the Rietgreb, all small lakes in the area northwest of Amsterdam which would become the 'Noorderkwartier', were drained and turned into valuable agricultural land. In 1556, the Count Van Egmond began the great undertaking of draining the Egmondmeer, quickly followed by others. Not all areas needed windmills. Between 1540 and 1565, the total area of land reclaimed in North and South Holland, Friesland, Groningen, Zeeland and North Brabant was 35,608 hectares of which only 1,349 hectares were drained meers. But then, between 1615 and 1640, the total was 25,513 hectares, of which no fewer than 19,060 hectares were obtained by draining meers.[32] To achieve this, drainage by windpower had to be further developed.

The power needed to drain any polder was determined by the volume of water and the height to which it had to be raised. The water had to be removed within a reasonable time to prevent the land becoming sodden. But marsh mills were limited in two ways. Obviously they needed wind, but they could be increased in size to take advantage of the wind only up to a certain capacity related to the strength of the materials available. The span of sails with wooden stocks was normally limited to a maximum of 30.48 m (100 ft.) which, in very favourable wind conditions, may have given about 20 effective horsepower at the scoopwheel. To increase output, the number of mills had to be increased. At Kinderdijk, one system can still be seen where the lift is limited and can be coped with by scoopwheels raising the water in a single stage (Figure 62). The

Figure 60. The Schermer polder was drained by these polder mills built on the North Holland style with winches for winding in the caps. This is a 'triple lift' set and the middle mill is now a museum (August 1990).

two polders, Overwaard and Nederwaard, have separate sets of mills. In one, there are eight smock mills with reed thatching, built in 1740. The second consists of a row of brick tower mills built in 1738. Both raise their water into canals which form storage basins from where it can be run off into the river.

The diameter of the scoopwheel and the length of the Archimedean screw limited the actual lift and, if a deeper polder had to be drained, then one mill had to lift the water to a second a little higher and sometimes the water had to be raised to a third higher still (Figure 63). These were the methods exploited in the seventeenth century and later. This system was probably invented by Simon Stevin (1548–1620) but is also associated with Jan Adrianjaensz Leeghwater (1575–1650) who was born in North Holland at De Rijp, a village then

Figure 61. The side view of this typical South Holland polder mill, the Blauw Mill, shows how the sails are inclined and how the tail poles are attached to the cap. To the right can be seen a winch for hauling boats out of the polder drain into the upper canal (March 1992).

Figure 62. Just a few of the mills at Kinderdijk showing the drainage channels from which the two sets of mills pump the water (May 1977).

Figure 63. The triple lift mills at the Tweemans polder (March 1990).

surrounded by marshes and lakes. He started life as a carpenter and millwright and ended as a renowned hydraulic engineer and dyke builder. He was responsible for draining the Beemster lake between 1608 and 1612. Proposals had been put forward to do this around 1572 and Leeghwater achieved the actual drainage in one year with twenty-six windmills. Because the depth of the lake was 3.06 m (10 ft.), one set of windmills lifted the water to a ring canal from where the second set raised it to the outlets. Although the Zuider Zee burst in and flooded the land so the work had to be done again, Leeghwater was involved with draining the Purmer in 1622, the Wormer in 1625 and the Heerhugoward in 1629.[33] In 1631, the States of Holland and West Friesland granted the town of Alkmaar a patent for draining the Schermer following Leeghwater's plans. The polder was divided into fourteen plots, each with its own drainage mill. These pumped the water into storage basins from where a triple lift system of mills and ring canals raised the water to the highest level. The mills were placed in twelve sets of three.[34] The number of windmills totalled fifty-two, which kept this area dry until they were superseded by electric pumps in 1928. Leeghwater also drew up proposals for draining the Haarlemmermeer in 1629 with 160 windmills,[35] but this scheme was never carried out.

One triple lift set remains the sole means of draining the Aarlanderveen polder, for in 1992 proposals to install electric pumps were rejected. Tests were carried out here between 1955 and 1958 to determine whether two of these mills, which are spaced at a distance of 286 m (300 yards), interfered with each other. It was found that, when the wind blew directly from one mill to the other, the loss of power in the second mill amounted to 12 per cent, corresponding to a drop in wind speed of 4 per cent. Evidently a Dutch windmill may cause a 'wind-shadow' which extends for a distance of more than ten times its own sail span, the figure on which the old millwrights worked.[36] As an illustration, the post mill at Drinkstone in Suffolk is situated on the top of a hill while a little lower is a smock mill. Although the post mill mound reached to a fair height, this did not shelter the smock mill, which worked perfectly with the wind from this direction until the post mill was started after which the smock mill would not run satisfactorily.[37] American wind engines with more solid wheels caused considerable wake rotation, thus limiting their efficiency.[38]

The performance of windmill drainage began to be studied systematically from the 1920s. Figures produced then and in the 1930s showed that only about 50 per cent of the horsepower generated at the windshaft was effective in raising water through frictional losses and through leakage. Traditional windmills began to turn in wind speeds of 5 to 6 m/sec. (11 to 12 m.p.h.) but at that rate the scoopwheel raised hardly any water. They needed a wind of over 8 m/sec. (18 m.p.h.) to be really in their element, when they would be turning at 75 to 85 'ends' (the numbers of tips of sails passing) and generating about 50 h.p. At a wind speed of about 10 m/sec. (22 m.p.h.) the sails would have to be reefed, for at 90 ends or slightly more there was a risk of the mill running away because the scoopwheel would be rotating too fast with the danger of cavitation. The windshaft horsepower generated with such winds would be over 60 and might rise to 90. In the Netherlands, wind speeds of 6 to 8 m/sec. (13.5 to 18 m.p.h.) occur on average 1,332 hours a year and 8 to 12 m/sec. (18 to 27 m.p.h.) for 1,339 hours. So the old windmills might work for around 2,671 hours in a year, or about a quarter of the time.[39]

The advent of the steam engine

The figures above will show why the Dutch were interested in using steam engines. The first to be erected was near Rotterdam in 1776 with a Newcomen engine, cylinder diameter 1.32 m (52 in.) and 1.83 m (6 ft.) stroke. The engineer was Jabez Carter Hornblower. Windmills pumped the water from the land into a storage basin that was the same level as low tide in the river but the tide rose 1.52 m (5 ft.). Therefore the engine had to work against different lifts according to the tide, so the power output had to vary. But the power developed in the steam cylinder remained virtually the same at every stroke, and the solution to this problem showed ingenuity even if it was not successful. The piston in the single steam cylinder was connected to three main beams side by side and two subsidiary ones. At the end of the beams were vertical lift pumps, some round and some square, while three smaller pumps were placed half-way between the outside ends of the main beams and their centres which had only half the stroke of the others. These pumps could be disconnected or joined on again individually while the engine was running in order to lessen the load as the tide rose, or to increase the amount of water pumped as the tide fell.[40] The comment was made about its performance:

> This plan has been found to be defective; for whenever one of the pumps is taken off or disjoined, the engine has too much power for some time; and commonly, when a pump is put on at the ebbing of the tide, she has too little power, and sometimes entirely stands for some time.[41]

In 1785, the Batavian Society ordered a 22 h.p. engine with a single lift pump from Boulton and Watt to drain the polder of Blydrop and Kool near Rotterdam. This was fitted with Watt's separate condenser so should have been more economical than the Newcomen one. Between then and 1807, three more engines with lift pumps were built for land drainage, which pumped out the water successfully but were heavy on fuel.[42] No more were tried until 1826 when three rotative beam engines driving scoopwheels were built at Arkelschendam. As these consumed 31 lb. of coal per h.p. per hour, it was thought that steam could not be used economically. The turning point came in 1837, when, in addition to thirty windmills, two 30 h.p. steam engines turning Archimedean screws were erected for draining the Zuidplas. These consumed 22 lb per h.p. per hour, which was considered reasonable in those days and helped to lead to the adoption of steam for the Haarlemmermeer reclamation.

Draining the Haarlemmermeer became a necessity because the prevailing southwesterly winds were causing waves which eroded the shores. In the autumn of 1836, severe storms inundated hundreds of acres of land and the town of Amsterdam itself was threatened. This instigated a serious proposal for using 114 windmills and in 1839 a law to drain the Haarlemmermeer was enacted, but in 1840 the plan of the engineers A. Lipkens, J. C. Beyerinck and Dr G. Simons began to be executed using three steam engines.[43] These schemes showed the limitations of windmills once the steam engine had been developed into a machine sufficiently powerful that one engine could replace many mills. The engineers submitted capital costs and annual running expenses for comparing 114 wind drainage mills with six 200 h.p. steam engines.

Windmills		Steam Engines	
	fl.		fl.
Capital costs (114 windmills; per windmill fl. 26,000)	2,964,000	Capital costs (6 steam engines, buildings & pumps; per system fl. 155,440)	932,640
Operating cost per annum		Operating costs per annum	
Interest & depreciation	148,200	Interest & depreciation	46,632
Maintenance & wages of 114 millers	74,100	Personnel (3 engineers, 18 stokers, & 12 labourers)	10,800
		Maintenance	15,000
		Oil, water, etc.	2,400
		Fuel	25,644
Total	222,300	Total	100,476[44]

Because the size of a windmill was limited by the materials from which it could be built at that time, a great many were needed, which increased initial capital expenditure as well as later depreciation allowances and wages. These outweighed the costs of coal so that steam engines were not only cheaper but could be made available whenever wanted.

Eventually, for the Haarlemmermeer scheme, three 350 h.p. engines were ordered, the first in 1843 to a novel design of compound engine which did not prove entirely successful even though it was based on the Cornish principle which had shown such economy in the tin mines there. This engine was demonstrated in November 1845 and orders were placed for two more. One of these, the Cruquius engine has been preserved. The compound cylinders in the middle of the engine are connected to a common crosshead from which beams radiate out to nine vertical lift pumps, a novel layout. The actual drainage of the 45,000 acres of the meer started on 7 June 1848 and on 1 July 1852 the announcement was made, 'The Lake is dry.'[45] The successful draining of the Haarlemmermeer showed that steam power could replace wind.

England

Windmills have drained land in many parts of England but their success has differed in each locality. In some their life was brief, but in the Norfolk Broads windmills were still being built in the opening years of the present century when they had vanished elsewhere.

The Somerset Levels

Around the periphery of the Somerset Levels lie Exmoor, the Mendips and the Blackdown Hills where the average annual rainfall is over 101.6 cm (40 in.). In central parts of the Levels themselves, the amount drops to 76.2 cm (30 in).[1] Of the nine highest recorded falls in twenty-four hours in the British Isles from 1865 to 1956, four have been in Somerset and all nine have been on the southern and western edges of the Levels' catchment area. The rivers may rise quickly and endanger the land. While some areas drained adequately by natural means, some remained rough peat up to the present and only now are being excavated for garden compost. Bleaden Level was reclaimed in 1613 and was noteworthy because the scheme included the first of only two known drainage windmills here. Quite why a windmill was necessary is difficult to understand because the fall from the level of the land to the low tide line is so great that a simple sluice to keep out high water is all that is required. The draining of Common Moor was completed by 1722 under the auspices of the first Parliamentary Enclosure Act in Somerset. A pump or engine, in all probability a windmill, raised the water and was demonstrated to a physician, Dr Calver Morris, in November 1723.[2] Even though the Levels are open to the prevailing southwesterly winds, the volume of water to be removed from the land was impossible for windpower and remained uneconomical for steam power until the second half of the nineteenth century.

The Humber estuary

There are various low-lying lands on either side of the rivers Ancholme, Trent, Ouse, Foulness and Hull, which drain into the Humber estuary. In the seventeenth century, about a dozen drainage mills were introduced into the South Yorkshire districts. Arthur Young saw one at Routh, northeast of Beverley, and, in 1770, published a picture of it (Figure 64) showing a form of wipmolen with the supporting structure left exposed and a large winding vane.[3] In 1764 a scheme was prepared by John Grundy and John Smeaton for the area near this mill, with five tower drainage mills, but this was never carried out. Instead the Holderness drain was dug, which drained the region naturally and dispensed with all the mills. In the same way, Smeaton drained the Adlingfleet Level in 1772 and Hatfield Chase in 1789, both to the south of the Humber estuary, by enlarging, straightening and lowering the drains and sluices which gave a natural drainage.[4] The other low-lying lands around the Humber could also be

Figure 64. Arthur Young's drawing of a drainage mill which was used near Routh, Yorkshire. It has an open trestle like a wipmolen and was winded by a vane (1770).

drained naturally so the windmills in this area disappeared with the introduction of better drainage schemes.

The Great Level of the Fens

As the history of the drainage of the Fens is very complicated and has been treated in detail elsewhere, the background is covered only briefly here.[5] Into the Wash flow the rivers Welland, Nene and the Great Ouse with its tributaries such as the Cam, Lark, Wissey and Nar. Where rivers and sea met, silt was deposited and formed banks and salt marshes which lay above low tide level. This process continues to the present day, so the Wash is gradually being filled up. The silt lands stretch roughly from Boston in the west, past Spalding to King's Lynn in the east. When embanked and the sea and rivers kept out, this land has great fertility, so it was quickly colonised. Villages such as Sutterton, Holbeach, Long Sutton, Wisbech or Outwell with their large churches show the wealth of the area. Most of the land could be drained by gravitation through sluices, provided that the outfalls of the rivers were kept clear, but parts needed the assistance of windmills which started to be built during the reign of Queen Elizabeth I.

To the south of the silt lands lie the black fens. Here fresh water from the rivers flooded the land because it could not escape to the sea. Aquatic plants flourished in the lakes and their deposits created marshes and eventually peat fens. Parts of the underlying subsoil are lower than low tide level, but the thickness of the peat used to be more than 12.19 m (40 ft.) in places, so its surface was higher than the silt lands. When Humphrey Bradley, from Brabant, was surveying the area in 1589, he found that the peat lands were sometimes at least 1.83 or 2.13 m (6 or 7 ft.) higher than low tide level and the tide rose about 5.18 m (17 ft.). Therefore he thought that these areas could be drained quite easily by gravitation provided that the water in the rivers could flow out to the sea freely at ebb tide.[6] Speculators were interested in the peat lands for their extraordinary fertility. Not only did they provide excellent pasture but, by burning the top surface and ploughing it in, the alkaline ash neutralised the acids in the peat and the lightness of the peat itself gave an excellent friable tilth. But if the level of acid water in the peat, the soak, rose, it would undo all the good of this treatment and so the water level had to be controlled.[7] In a successful year, agricultural rewards were great and oats were a prolific crop.

In the 1620s, a group of 'Adventurers' headed by the fourth Earl of Bedford proposed to drain the Fens. Another Dutchman, Cornelius Vermuyden, recommended straightening drains and rivers to shorten their courses to the sea and improve their flow, so that they would scour out deeper channels thus enabling water to escape more quickly. The result of his work can be seen today in the Old Bedford River, cut after the 1630 Act; the tidal New Bedford River, or Hundred Foot River, cut as a result of the 1650 Pretended Act, which stretch from Earith to Denver Sluice; and many other channels forming a network over the area. The North, Middle and South Levels of the Bedford Level were created at this time to administer those lands which had been drained, names which still denote current drainage areas. Most of these new channels were suitable for navigation and more of the Fen rivers were improved during the eighteenth century. While the Great Ouse River had been navigable for centuries, the first Act for making the Cam navigable to Cambridge was passed in 1702, the Lark in 1700, the Nene in 1714, and the Nar in 1751, to give a few examples.[8] Kinderley wrote in 1751:

> This River [Ouse], by its Situation, and having so many navigable Rivers falling into it from eight several Counties, does therefore afford a great Advantage to Trade and Commerce, since hereby two Cities and several great Towns are therein served; as *Peterborough*, *Ely*, *Stamford*, *Bedford*, *St Ives*, *Huntingdon*, *St Neots*, *Northampton*, *Cambridge*, *Bury St Edmunds*, *Thetford*, etc. with all sorts of heavy commodities from *Lyn*; as Coals and Salt (from *Newcastle*), Deals, Fir, Timber, Iron, Pitch, and Tar (from *Sweden* and *Norway*), and Wine (from *Lisbon* and *Oporto*) thither imported; and from these Parts great Quantities of Wheat, Rye, Cole-seed, Oats, Barley are brought down these Rivers, whereby a great foreign and inland Trade is carried on, and the Breed of Seamen is increased.[9]

The Fens had a system of internal transport by water for heavy goods which was better than that of almost any other district in Britain at this time, and this was one of the factors lying behind the land drainage improvements.

But soon after the Duke of Bedford's attempts at improvements, problems both in the navigations and on the drained lands reappeared. What was not appreciated for very many years was the extent to which peat would shrink when drained. It was realised that the peat would become more compact as the water was removed but, once this had been done, it was thought that the land would become stable. Some people thought that the beds of the rivers or ditches became higher but, in 1649, Dodson commented,

> It is not your Dikes bottoms which rise, but your Grounds which sink, and become much better; therefore when your Grounds are thus sunk with lying dry, bottom then your Dikes two foot, and your Dikes will hold good for many years.[10]

However, we now know that Dodson was mistaken: for peat, being vegetable matter, wastes away through bacteriological action when drained and in effect disappears completely. The land shrinks quickly through both causes but slows down the closer the surface approaches the new water level. It becomes sodden as it nears the new level, so farmers then want the level of the water in the drains lowered again. This continues until all the peat has disappeared. Whittlesey Mere is a famous example; near there, in 1852, the Holme Post was driven into the ground until its top was level with the surface soon after the mere was drained. Today at least 4.57 m (15 ft.) protrude above the ground in the wood where it is situated and this would be greater had the area been under continuous cultivation. Furthermore, peat had been dug for the embankments. Not only was it porous but it wasted when dry so that the banks became useless. At the time of Vermuyden, the complaint was made about peat for banking:

> for being dry it is so spungy that it will both Burn and Swim; and is so hollow that Banke which is this yeare large and firme to the eye, in foure or five yeares will shrinke to lesse than halfe the proportion which it had at the first making.[11]

Indeed, at the end of the eighteenth century it was said that more water would soak through the banks in the night than the windmills pumped out during the day, even when there was a good wind.[12]

The state of the black fens became parlous during the early part of the eighteenth century. Navigation had to be carried out on rivers that were barely passable and the land became more and more sodden. Windmills were employed for drainage but often proved useless through lack of wind. One answer was to improve the outfalls of the rivers. There were many years of arguments but, on the Cam and Great Ouse system, the Eau Brink Cut above King's Lynn was opened on 31 July 1821. The Nene Outfall Cut followed on 4 June 1830.[13] While the Nene Outfall, with a new North Level Main Drain, gave the North Level a natural drainage which dispensed with the need for windmills, the lands further south were not so fortunate. Large areas in the Middle and South Levels could not be drained by gravity but benefited through flood levels in the rivers being considerably reduced. In 1791, Golborne pointed out the advantages of lowering water levels in the rivers.

> The advantages given to a water engine in the time of distress, or in high floods, by lowering its head of water only two feet, and the pressure and danger of these two feet taken off their banks, may be more generally known than I can pretend to describe, particularly to those that are more universally acquainted with such engines. In the last winter floods, the loosing of only two inches of water against several districts, was looked upon as auspicious, and was testified by visible signs of joy in the counternance of many worthy and industrious farmers, who were deeply interested in the preservation of such districts from drowning.[14]

The realisation that the outfalls could be improved no further and that no more relief could be gained in that way made farmers in these parts of the Fens realise that they had to maintain mechanical pumps and so they turned more and more from windmills to steam engines.

The introduction of windmills

It is impossible to say when windmills for drainage were introduced to the Fens (Figure 65). Monasteries were concerned with improving the lands they owned in that region and would have had the resources and organisation to build windmills. In the records of Courts of Sewers for 1395, it was decided

> that the Towne of Spalding ought to repayre and heighten the said banks in Spalding next Welland water . . . from the Abbots mill [of Crowland] unto Spalding Drove . . . and that Thomas the Abbot of Croiland and the Convent of the same place and their ought to repair the same bank next Welland water from the said messuage of Wm. Kellod unto Dole mill dike.[15]

Whether these mills were for grinding corn or raising water cannot be established. In the records of the monastic estates during the dissolution of the monasteries by Henry VIII between 1535 and 1540, there are references to a *molendum* at Thorney and another at Ramsey[16] which, from their situation, cannot have been watermills, but their function is unknown. Windpowered drainage engines are thought to have appeared after this period.

One difficulty is to know what term would have been used to describe a windmill that pumped water. 'Engine' is the English term used during the late sixteenth century and in official documents it was retained until well into the eighteenth.[17] Yet by the beginning of the seventeenth century, the words engine and mill were interchangeable for describing a water-raising windmill.[18] If 'windmill' were one term, then a windmill shown on Hayward's Map of the Great Level in 1604 near Fosdyke on the sea-bank might be the one mentioned by the Commissioners of Sewers in the parts of Holland (Lincolnshire) in 1555 who complained that 'the see [bank] bilongyng to Algerkyrk and Fosdyk from the wyndemill unto Cromer Hyrn is in decay'.[19] A site inspection revealed remains of a windmill beside the old sea-bank with at least two ditches leading up to it, so it must have been used for pumping water and thus was one of the earliest in the Fens as well as in Britain.

Windmills were employed in the Fens during the reign of Elizabeth I, when various people obtained patents for 'engines to drawe waters above their naturall Levill and to drayne waterishe and moorishe grounds'.[20] In 1578, Sir Thomas Goldinge asked for a patent for he claimed to be able by his engine 'to drye all places drowned or under water'.[21] Daniel Houghesetter had an engine which would raise 'waters from anye place whatsoever from low to high',[22] and William Mostart applied in 1592 for the privilege alone 'faire des engins pour secher les mares scayes et fennes basses'.[23] The most important early speculator was Peter Morrice because we know that, although his petition to the Queen contained no description, it was connected with drainage windmills. In 1575, he petitioned 'Mr Fraunces Walshingham, principal Secretary to her excellent majesty':

Figure 65. W. Blith's drawing of a drainage mill must be the earliest one printed in England but there is no provision for winding the sails (1652).

Peter Morrice yor poore orator hath to his extreme charge and with like payne travell and industrie endevored to make divers engins and instruments by motion whereof running streames and springes may be drawen farr higher than their naturall levills or course and also dead waters very likely to be drayned from the depths into other passages whereby the

> ground under them will prove firm and so moche the more fertill which thing if it comes to perfection will be a most high and great comoditie to the subjects of all partes of this realme . . . that it would please her M'gesty to grant unto him the seid patent to authorise him and his only to make and sel the same engines . . . forbiddinge all other . . . to make or use the same within thy realme unless it be by consent of the said Peter Morrice . . . for the term of 21 years.[24]

The style and form of this petition are similar to later patents except that it is for twenty-one years and not fourteen. Morrice failed to derive any benefit and the rights were sold to George Carleton in 1580,[25] who proposed to set up his engines and drain grounds in the south of Ravens Dyke near the towns of Whaplode, Holbeach and Fleet.[26] The objections raised were typical of people fearing that an improvement in one area would cause deterioration in another, a theme that recurs frequently in fen drainage history. It was felt that these 'engines were dangerous to the countrie, of great charge and small performance, and experience hath well taught so', and permission was not granted.[27] The phrasing of this rejection suggests that similar windmills had been tried elsewhere and that Carleton's were not the first.

Carleton had wanted to place one of his engines on the sea-bank between Gedney and Sutton Goates, an area now far inland. It would have been necessary to cut the bank to make the discharge sluice or 'goat' but it was claimed that this would seriously weaken the bank. Also, it was feared that, if he made his sluice 30.48 cm (1 ft.) lower than the existing three, drainage through them might be adversely affected.[28] It was felt that Carleton ought to be content with existing drainage works and not try to improve his own lands at the expense of others. This was a view often repeated against windmills and even steam engines, so that the balance of the whole drainage system might not be upset by the individual. In Carleton's case, it was felt that

> the most part of all the other lowe grounds will be convenyently drayned by the old anncient and accustomed draynes, without any innovation, yf the same old draines were effectually repaired and amended.[29]

However, Carleton was able to erect some engines for one came to an unfortunate end in 1584.

> Holbech having an Ingen for them erected and sett upon the sea, to delyver their water had no greate dannger thereby. When the same standing but VIII tydes in chiefe of wynter took their ryver downe, and laid their goate drie. And as no danger; so no chardge, because the same was geven by Mr Carleton at his owne coste, as like is now offered for Sutton and Gedney by such as shall draine that waie. Indeede that Ingyn had no contynuance because the fourth daie after the same was finished, and in her works, it was by the scheme of some wicked mynde overthrowen . . . What Ingen sover that hath such handlinge cannot longe contynue but if such

neighbourhood to the same had bene in Holland as Mr Carleton fyndeth about Wisbech, Holland Ingyns might have prospered.[30]

The person who destroyed this mill, William Stowe, was eventually apprehended and brought to trial. It appeared that after the church service on Christmas Day, Stowe complained to the Dikereeves that they allowed windmills to be erected on the river banks. He left the church exclaiming that he would fight against these intrusions but received no support from anyone else. Someone cut one of the beams supporting the mill almost in two so that, two days later when the miller climbed into the top to oil it before starting, 'the north east wind blew from the sea and brake the said beame in the place so cutt and threwe down bothe man and fframe'.[31] The collapse of this mill when one beam was sawn through shows it must have been a wipmolen rather than a smock mill. While early maps like Haywards show Fen mills as post mills, later ones about which there is definite evidence were all smock mills. The account of the destruction of Carleton's mill makes it clear that opposition was not general but confined to isolated people and the Privy Council seems to have looked favourably on windmills as a means of improving the Fens.[32] But Carleton soon found out their great drawback: that the wind to work them did not always blow at the right time. Like many other fen farmers after him, he waited in vain for a wind while his 'groundes called Coldham, where he hath used all his pollicye and ingens, is at this Daie . . . drowned as ever it was'.[33]

In spite of the ineffectiveness of windmills, their numbers increased and spread across the whole of the Fens (Figure 66). While eventually those on the siltlands in the north were to become redundant, the black fens relied more and more on them. In view of what happened later, after the establishment of the Bedford Level, it is necessary to take note of those mills which had been built in the southern half of the region. Richard Atkyns, in a survey of the Fens in 1604, mentions certain grounds of Sir William Hindes at Over 'where ther is an Ingin or Mill placed to raise water – and not farr fro there another mill for the town of Over, both serve to good purpose and empty the water into a dike which falleth into Willingham mere'.[34] Hayward's Map of the Fens in the same year shows a number of windmills scattered throughout the area. Casaubon saw a mill when he was travelling through the Fens with the Bishop of Ely in September 1611. They lost their way while returning from Wisbech and went to a windmill where they found a boy who showed them the right path to a ford. Here the Bishop's horse reared up and the Bishop fell off but came to no harm, although Casaubon took some hours to recover from the fright.[35] There was a mill or engine built at Leverington before 1617[36] and no doubt others in the area which later came under the jurisdiction of the Bedford Level Corporation.

An Act which received the Royal Assent on 13 September 1660 confirmed the Bedford Level Corporation and judged the lands drained.[37] The Corporation was empowered to raise money for the upkeep of the system by levying taxes on land within its boundaries. But soon the Corporation found that it could barely maintain the existing works, let alone initiate new ones as conditions deteriorated. Everywhere there was more work to be done than there was money available. For example, in 1701, Mr John Reynolds, Steward to the Duke of Bedford, informed the Corporation

> that the money allotted this yeare for the Workes of the North Levell is not sufficient to putt the Workes of the said Levell in a defencible condicion against any ordinary flood that may happen.[38]

Figure 66. This old photograph of Isleham Mill near Ely shows how Fen drainage mills must have appeared (Reid Collection).

Sometimes the Corporation was forced to employ the doubtful expedient of using funds that had been allocated for somewhere else, or to complete work on the credit of the following year's taxes.[39]

As conditions deteriorated, farmers looked after their own interests and resorted to artificial means for draining their lands. At a meeting of the Bedford Level Corporation in London in May 1663, there was an application from John Trafford to drain 500 acres of fen at Tydd St Maries by an engine into the Clough's Cross Drain. Permission was granted 'so long as the same be not prejudiciall to the venting of the waters of the Great Levell of the ffens called Bedford Levell'.[40] In later Minutes, there are other references to engines or mills. Sometimes permission was being sought to erect an engine,[41] but more usually there was a complaint against a particular mill. Often there was a certain amount of self-interest in the action when the mill concerned threatened the Corporation's lands. Such a case in 1680 was over mills in Coldham and Waldersea, 'whereby the Adventureres grounds there abouts were much dampnified'.[42] The offending mills were not ordered to be pulled down, but a bank was built to prevent the water thrown out by the mills running into the Adventurers' grounds of Lades and Crooks.[43] The result was that 'by reason of the said bank, the said inhabitants were in a very badd condition having noo other ways at present to vent their mill waters whereby theyr grounds would become useless and the present crops thereon bee utterly spoiled'.[44] They asked that they might be able to cut a watercourse through the bank as a temporary expedient until they could find some other way of draining before the following spring.

Mills continued to be set up without any supervision and the nuisance they caused greatly increased. In 1693, Lord Torrington proposed setting up two mills in the Whittlesey area that threatened the Adventurers' lands,[45] and there were a great many others in that district. One of the Corporation's officers, Mr Bourne, was ordered to find out how many mills there were on Moore's Drain and Bevill's River, what size and to whom they belonged.[46] Enclosure of some drier parts of the Fens began at this time which, as well as the increasing numbers of mills, aroused opposition, perhaps because land was drained from which poor people had been able to make some sort of living by fen pursuits such as wildfowling, cutting peat for fuel and grazing a few cattle. Outside the actual Bedford Level in 1698, great harm and destruction was 'committed by divers desparate and malicious persons, that have destroyed in a great measure the works of draining in Deeping Level . . . under cover and pretence of football playing'.[47] Houses, buildings, mills, banks and other drainage works were destroyed, and the Bedford Level Corporation was warned that the rioters intended to meet at Coates Green near Whittlesey to pull down the mills and cut the banks, again under pretence of 'Football Play and other sports'.[48]

Perhaps as a result of these riots, the Bedford Level Corporation took active measures against horse and windmills and, acting as a Court of Sewers in three sessions in 1700, 1701 and 1708, fined, or ordered to be stopped up, forty-four windmills and thirty-seven horse mills.[49] The indictment against John Walsham in 1700 was typical.

> Wherefore it was the same day presented by the same jury that John Walsham, gent, for working his engine or mill which through mudd and stuffe into the 20ty. ffoote Draine called Mooree Drayne so Mudd and stuffe choaked and stopped the currant of the water of the said drayne in its passage to the River Neane and therefore prayed that the said John Walsham may be enjoyned by the authority of this Court not to worke the said engine or mill on this side and before the – day of – next and in default thereof to forfeite to the King's Majesty's use the sum of 10/– if he shall doe the same, ordered the same to be stopped accordingly.[50]

He was brought before the Court on all three occasions, and half a dozen others appeared twice, although the fine was increased to £100 in 1701.[51] None of these mills was ordered to be removed, for this was beyond the powers of the Corporation. All were charged with causing an obstruction in the rivers or drains, and therefore had to pay a fine or cease working, but windmills had become a necessity by this time.

Windmills could be used by anyone provided there were no complaints, which was made clear in two court cases. That in 1703 against Sylas Titus was not to try whether mill owners had a right to work and use mills,[52] for it was acknowledged that they were free to build and erect them, but whether owners could use machines which damaged or drowned their neighbours' lands while draining their own.[53] The point was that nobody should 'erect or make any works for draining, at their own wills and pleasures, their own lands to the damage of public draining, and the prejudice of other lands adjacent upon any pretence whatsoever'.[54] People should be content with the usual way of draining provided by the Conservators of the Bedford Level. If they caused a public nuisance by blocking up drains with silt from their mills, fines were imposed.[55]

Where there was no nuisance, the Bedford Level Corporation could do nothing which was borne out in the case against a Mr Hyde. In 1699 he applied for permission to build a mill to throw the water from Sutton St Edmund into the Shire Drain. Although this area is not in the Bedford Level, the Corporation was responsible for the drain because it carried away the waters of the North Level. A committee was appointed to view the place and permission was to be given if its report decided that the mill would not obstruct the water flowing from the North Level.[56] But there must have been some other opposition, for the Bedford Level Corporation started to prepare a case for Counsel.[57] After alternately being granted permission to build a mill in various places and then having it withdrawn,[58] Hyde decided to go ahead. Officers of the North Level ordered the workmen to stop erecting the mill, but they continued.[59] The Corporation was determined to stop this mill and sought advice from Counsel,[60] while orders were given for the water level to be carefully observed. One of the Corporation's officers, Mr Le Pla, claimed that the mill did raise the water,[61] but he could not have had time to take accurate measurements. Then the affair died down. In London, it was decided to consider the matter another time,[62] while the Ely Proceedings show that Hyde wanted witnesses appointed from both sides to view the Shire Drain when his mill was working.[63] Clearly the Shire Drain was well able to 'bear the Waters of St. Edmund',[64] so the windmill remained because it did not cause a nuisance.

There were further disturbances against windmills,[65] but the riots died down and the number of windmills increased. In 1721, Richard Saffery received permission direct from the Corporation to erect a mill on the north bank of the New Bedford River between the Wash (the lands onto which floodwaters could be discharged) and the Old Bedford River so he could drain his lands better.[66] In 1727, the inhabitants of Cottenham were allowed to build a mill or engine on Chear Fen bank between Sir Roger Jenyns's mill and Twenty Pence Ferry to drain their water into the River Ouse, provided they made good any damage they should do to the Corporation's works.[67] Sir Roger Jenyns was a Conservator who had defended the Corporation against Sylas Titus, but he possessed a mill at that time or soon afterwards. It was not much use, for when Badeslade took a boat at Stretham Ferry and 'went to Sir *Roger Jennyn's* Engine Mile, which is placed upon a great drain belonging to *Cottenham*; those very lands which this Engine is intended to drain, (and which was heretofore rich Meadow) we found drowned'.[68]

The Bedford Level found itself facing increasing indebtedness as it struggled to keep the waterways and banks in repair. In the South Level, the situation was probably made worse when Denver Sluice was undermined and blown up by the tides in 1713, which affected the lands bordering the River Cam. There was no money to repair it, and it was not rebuilt until 1746. Around 1725 there was a series of very wet years and the extra expenses entailed by many great floods in the two years before March 1726 ran the Corporation into heavy debt.[69] The country under its care suffered calamitously; the whole area was under water most of that time as the outfalls were obstructed.[70] Badeslade states that were it 'not for a great number of Landholders throwing the Fen Waters over Banks into the River and Drains, by Engines made at their own Expense, they say, the whole Body of the Fens would become unprofitable; and Taxes enough could not be raised to maintain the Works, and to pay Salaries'.[71]

The next stage in the development of drainage with windmills started when the Haddenham Level obtained a drainage Act. The evidence given before Parliament by John Clark, James Fortry, Mr Walker and John Kent stressed that the Level had long been drowned and that the only remedy was drainage by mills or engines.[72] But, in the correspondence with the Bedford Level Corporation, mills or engines were never mentioned. In April 1726, the inhabitants of Haddenham sent a petition to the Ely Conservators complaining that the bank of the Hundred Foot River was too low, for the floods had come over it twice during the last winter and greatly damaged their land. They asked that the banks should be raised and some way found to improve the outfalls to the sea.[73] In the following April, a similar petition was presented, this time complaining about the north bank of the Ouse between the Hermitage and Stretham Ferry, and also about the state of some of the ditches.[74] From this time to 1731, the Corporation spent over £400 on the banks of the Haddenham area but, in the meantime, permission had been given by the London Committee of the Bedford Level Corporation for the inhabitants of Haddenham to drain their lands 'at their own expense'.[75] No objection was raised so an Act was promoted and passed.[76]

Immediately afterwards, this was followed by the Whittlesey and Waterbeach areas. Before a Committee of the House of Commons in February 1728, Mr George Claxon said that

> the low Lands, and Fen Grounds, in the Township of *Whittlesey*, containing Seven thousand Acres, or there abouts, have until of late, been very good, and yielded great Profit; but the same, for the Space of Seven Years last past, and upwards, have been so often drowned; and surrounded with Waters, that the same have yielded little or no Profit to the Owners: That he has Fifteen Acres in the said low Lands, but has made nothing thereof for these Three Years; but had formerly Six Pounds a Year: That he believes, that the cause of the said low Lands being drowned is the want of proper inland Drains, and embanking the old Drains, and raising proper Engines, for throwing the Waters into the common Rivers.[77]

The Waterbeach Level also sent a petition to Parliament[78] but their Bill was never read when the Whittlesey Bill failed to pass beyond the Committee stage.[79] It is not known whether this was the result of opposition or whether the initiators had gained what they wanted, for about £600 was spent by the Corporation on the banks of Waterbeach Level before 1732.[80] Another reason may have been the promise of a general scheme for the Fens, as various ones were suggested during the next few years.[81] While the years 1729 to 1731 did see an improvement in conditions,[82] it is suspected that there must have been some opposition or legal hitch because the next successful drainage Bill was not passed until 1738 for the very small area of Redmore, Waterden and Cawdle Fens near Ely,[83] followed by a second successful application from the Waterbeach Level in 1741.[84] From then on, many other areas procured their own individual drainage Acts.

These Acts authorised the establishment of Boards of Commissioners, to be responsible for the internal drainage of their districts. They were empowered 'to make such Cuts, Drains, Damms and Outlets, through the said Fens and Low Grounds . . . and to make and erect such Works and Engines thereupon, for draining and conveying the Waters from the same'.[85] In order to maintain all these works, they could levy an annual tax, usually about a shilling per acre. This was made a statutory demand, based on ownership of the land, with powers of seizure for non-payment. There were other sections concerning the cleansing of drains, with penalties laid down against anyone who should destroy any of the mills or engines or obstruct any of the drains or other works.[86] With such an Act, an area could obtain legal protection for work carried out, and compel the occupiers of the land to subscribe towards such work, so it was at last possible to have as competent a drainage system as the limits of windpower allowed.

The mills and millers

Fen windmills were nearly always built of wood. Only one very small example has been preserved, at Wicken Fen (Figure 67). A larger one survived at Soham Mere until 1948, when it was classed as a dangerous structure by Cambridgeshire County Council and ordered to be demolished. It could not be pulled over by a tractor nor could it be blown up by gunpowder and finally eight charges of gelignite destroyed it.[87] Usually fen mills had four sails, although there was one with six at Daintree, and Daniel Defoe described a model of one with twelve, but he did not see this working.[88] The sails of the largest mills were 29.26 m (96 ft.) from tip to tip while on one near March, they were each about 10.97 m (36 ft.) long and 2.28 m (7 ft. 6 in.) broad.[89] They were always of the common type with canvas and never any type of patent sail with a fantail. This may have been because the millers had more time in the absence of duties such as grinding corn. The mills were turned by tail poles and winch with ten or twelve anchor posts around the base.[90]

Some mills were built on piles driven through the peat into the clay. Others had brick bases with smock mills built on top. These bases might extend as high as the first floor; examples can be seen at Wiggenhall St Mary Magdalen, and the double lift mills at Nordelph where the upper mill has part of the wooden smock mill still surviving. At Dyke the framing of a marsh mill from Deeping Fen has been recently restored. The mill was moved in 1845 and converted into a corn mill, so it now shows little trace of its former usage. Fen mills were covered with vertical boards. Normally there was a wooden scoop-wheel housing too, although a scoopwheel survives in the open at a mill which probably had a brick tower at Amber Hill in Lincolnshire (Figure 68). The scoopwheels were usually outside the mill bodies. Size was determined by the anticipated amount of water to be removed and the size of the mill itself. In 1741 at Deeping Fen, two wheels which were then considered large were built 4.88 m (16 ft.) in diameter and 33 cm (13 in.) wide for draining water off the fen into the main outfall, Vernatt's drain.[91] The mill at Soham Mere, built in

1867, had a scoopwheel 7.62 m (25 ft.) in diameter, which was probably bigger than most.[92]

Because the mills had to be built well away from any obstructions in order to catch the wind,[93] the millers were housed on site and often had their families living with them, although the space inside was restricted by the gearwheels

Figure 67. The last surviving Fen windmill is this small one moved from Adventurer's Fen and now preserved at Wicken Fen where it is sometimes used to pump water onto the land to keep the marsh wet! (March 1966).

Figure 68. At Amber Hill, Lincolnshire, by the stump of a brick tower mill, is the last windpowered scoopwheel which survives in its original location in the Fens (September 1990).

and shafting. The vertical drive shaft had to be in the centre of the mill and so, therefore, did the gearwheels driving the scoopwheel. These might be shielded from the living section by a tarpaulin drawn from side to side. There were no windows, but there would be a fireplace, although this had no chimney; the smoke from the fire had to make its way out through cracks in the top of the mill where it acted as a preservative. In one dark corner there would be a bed and a wooden chest. Such was the dwelling in which a family lived permanently,[94] but it could be as good as one for an agricultural labourer at that time.

Two men would usually work one mill, or perhaps one man with the assistance of his family. They would be allowed to live rent-free in the mill, even during the summer when the mills were 'set down'. There might be various repairs which had to be carried out during this period for which the millers received pay, or sometimes the mill would have to be worked in an emergency during a wet summer. Otherwise they received a wage only during the winter, when the mills were actually working. The variability of their earnings is shown below by the totals of wages at each windmill in the Haddenham Level. Usually there were two men, but the Sutton mill may have been attended by a family.

Year	Highbridge	Horseshoe	Ferry	Lazier	Sutton
	£. s.d.	£. s.d.	£. s.d.	£. s.d.	£. s.d.
1739–40	7.18.0	7.19.0	7.17.0	5.19.0	7.13. 5
1740–1	7. 0.0	6.18.0	6.10.0	8. 1.0	4. 9.11
1743–4	4.13.0	5. 9.6	4. 0.6	4.12.0	2.10. 6
1744–5	14. 8.0	16.16.0	14. 8.0	14. 8.0	8. 6. 0[95]

There was a further bounty of free fuel and the Haddenham Level Account Books show that turf or peat was sent to the mills. Later coals were given instead. This tradition survived until well into the present century, for the stokers and superintendents of the steam engines also received free coal. At the end of the eighteenth century, millers received 10s.6d. a week with coals,[96] and were ordered to work whenever there was a strong enough wind, even during the night. A figure in 1848 was two shillings a day.[97] Normally they would be responsible only for working the mill, although they might be employed casually on other odd jobs when the mills were set down. George Carleton expected his miller to do considerably more and was going to give him a separate house, to be

> builded under Sutton Sea banke uppon some convenyent place for a man to dwell in, to keepe the Engen going in wynter, to looke to the Doares of all the Sleuces and goates as well Gedney Sutton as the new workes, and to spend his whole yere else in scouringe the channell; clensing the ffynnes and creeks, repayringe the two creestees upon the marshes, and taking the

> charges of coale, tymber or whatsoever else shall be upon that wharfe unladen by shippe or carte.[98]

Such low and variable wages must have caused the men to look for alternative income and, with the fens all round them, there were traditional ways of obtaining a living. One was by fishing; and apparently the millers employed by the Waterbeach Level in 1775 had not been attending to their work, so the order was issued that 'no one concerned in that business shall for the future fish during their continuance in our service'. The temptation must have been too strong for the instruction had to be reissued in 1793 and the Commissioners called on the Receiver to enforce it![99] In this Level, a special committee of three commissioners was appointed to supervise the work of the millers. In the Littleport and Downham District, which had seventy-five windmills, in 1810 Jacob Badger was employed for a year as superintendent over the mills and the district at a salary of £60 per annum. John Hare was appointed 'to do the Millwright Business of this District for one year with a salary of Sixty Guineas per Annum and Four Guineas for House Rent and allow his boy Eight shillings per week'.[100] While most Levels had their own superintendents, the position of millwright was probably occasioned by the large number of mills in this district.

Capital costs of drainage mills are difficult to establish because the value of money altered over the centuries and generally no indication of size is given. In 1665, Dodson wrote that the large Dutch mills like those 'on the Bempster or Skermer' cost about £600 sterling.[101] A drainage mill erected at Tydd St Giles in 1693 cost £450[102] and this was the price for another in 1710, with a 4.88 m (16 ft.) diameter wheel, dipping 1.22 m (4 ft.) and raising the water 1.22 m (4 ft.)[103] At the end of the eighteenth century, one of the larger windmills would cost more than £1,000. The Waterbeach Level paid £1,300 in 1814 for their new double lift mill.[104] At about the same time, Arthur Young mentioned one at Wilmington for £1,400 and two at Wisbech, which were rated at 9 h.p., for £1,200 each. Young also gave prices for mills in Burnt Fen costing £600 and elsewhere small mills costing £80.[105]

The amount of water that had to be pumped out varied greatly from year to year and from Level to Level. Some areas were completely isolated from higher land so were concerned with only rain, snow or water which soaked through the banks. Into other areas water ran down from the uplands but in some of these catchwater drains were dug to cut off the water from the higher land. The amount that fell as rain can be estimated from later records, but few figures were kept before 1800. John Rennie had to admit, in his report about Deeping Fen in 1820, that he could give

> no correct account of the Total quantity of rain that falls on the surface averagely in the year, on Deeping Fen, but from experiments made in other places similarly situated, it cannot be reckoned at less than twenty Inches,

and of this quantity, instances can be produced of two inches falling in the course of one day.[106]

The total in any individual year varied between 30.48 and 76.2 cm (12 and 30 in.), but an average from 1840 to 1925 at the Stretham engine, Waterbeach Level, works out at 53 cm (20.88 in.) per annum. This is a little lower than figures given by G.D. Dempsey for 1836 to 1843, which averaged 67.4 cm (26.61 in.). Dempsey also produced figures to show how much water might remain to be pumped out after allowing for evaporation, which varied according to the amount of sun, the state of the soil and vegetation.

Month	Total falling (inches)	Evaporated (inches)	Remaining (inches)	Waterbeach Records, averages of rainfall 1840–1925 (inches)
January	1.847	0.540	1.307	1.38
February	1.971	0.424	1.547	1.07
March	1.617	0.540	1.077	1.26
April	1.456	1.150	0.306	1.36
May	1.856	1.748	0.108	1.56
June	2.213	2.174	0.039	1.90
July	2.287	2.245	0.024	2.46
August	2.427	2.391	0.036	2.37
September	2.639	2.270	0.369	1.96
October	2.823	1.423	1.400	2.22
November	3.837	0.579	3.258	1.72
December	1.641	0.164	1.805	1.62
Totals	26.614	15.320	11.294	20.88[107]

It will be noticed, especially from the Waterbeach Level records, how the greater part of the rain fell in the summer months when most of it would be evaporated. The autumn too, tended to be wet when the rain would make up the balance lost through evaporation during the summer months.

The length of time the windmills functioned depended on the weather. They were seldom worked for less than six months and often for eight or nine,[108] being out of use during the summer. Usually in summer there was not enough wind to drive them, although in 1789 there was sufficient to keep the two windmills of Middle Fen near Ely at work all the season, for the summer was very wet. William Swansborough, an experienced fen engineer of the early nineteenth century, reckoned that windmills worked on an average one day in five,[109] for a considerable wind velocity was necessary before the mills could work effectively. Joseph Nickalls stated that 'the force of wind, acting equal to the strength of four horses was unable to open the water-gate door'.[110] Rennie commented in his report on Deeping Fen in 1818 that even with good breezes the quantity of water raised by the windmills did not clear the fen and prevent injury to the land.[111]

Analysis of Wind Velocity Compiled from Hourly Tabulations, Meteorological Station, Mildenhall, 1950–1959.

Mean wind speed (m/sec.)	Beaufort scale, equivalent	Total hours, in percentages
0.3		2.7
0.3–1.5	1	19.2
1.6–3.3	2	24.5
3.4–5.4	3	27.4 Moderate Breeze
5.5–7.9	4	20.3 Moderate Breeze
8.0–10.7	5	4.1 Strong Breeze
10.8–13.8	6 and 7	1.0 Strong Breeze to Moderate Gale
13.9–17.1	8 onwards	0.1 Fresh Gale, etc.[112]

The traditional type of windmill will work with wind speeds from 6 to 12 m/sec. (12 to 27 m.p.h.), with the ideal between 8 and 10 m/sec. (18 to 22 m.p.h.). Based on the wind speed figures given above, a windmill in the Fens would work for 1,778 hours with winds between 6 and 8 m/sec. (12 to 18 m.p.h.) and 450 hours with winds between 8 and 12 m/sec. (18 to 27 m.p.h.). While the Fen total of 2,228 hours compares with 2,671 for Holland, the figures for the more effective wind speeds are 450 hours for England and 1,339 hours for Holland. Therefore the time the English mills could work effectively was much more limited than their Dutch counterparts and goes a long way to explain why windmill drainage in the Fens was so much more problematical. This becomes clearer when the running hours of the Stretham engine are examined. 1919 was an exceptional year, because a bank collapsed in a flood and the engine ran continuously for 40 days and nights, giving a total of 2,707½ hours that year. No windmill could have matched that! 1883 was another bad year when the engine was pumping for 1,776 hours; but it ran for only 62 hours in 1898. Averages between 1839 and 1925 were 769 hours running and 54 cm (21.21 in.) rainfall. Presumably the engine was pumping effectively, so that this should be compared with the 450 hours when windmills would have worked reasonably and shows how steam engines could guarantee a much better drainage in the Fens.

There are few figures available of windmill performance. The best, in 1814, could lift 2,000 cubic feet of water per minute when the difference in levels did not exceed 1.67 m (5 ft. 6 in.),[113] and a large one would drain about 1,000 acres. Other figures indicate that a mill might drain more, for Haddenham Level with 6,000 to 7,000 acres had five mills. The Waterbeach Level with nearly 6,000 acres originally had three windmills, which perhaps explains why drainage here was poor until a steam engine was installed. Tycho Wing gave the number of windmills on lands draining by the River Nene and the acreage in a letter in 1820:

The North Level	48,000 acres and has	43 Engines
Wisbech North side, Tydd & Newton & Leverington	17,756	12
Sutton St. Edmunds	4,040	2
Waldersea & Begdale	8,000	4
	77,796	61[114]

Averaging the preceding ten years, he estimated that the 43 large mills each cost £80 per annum to work and repair while the 17 small mills each cost £25 per annum, compared with the annual expense of £200 for the steam engine at Sutton St Edmund. To secure better drainage, Labelye thought that the number of mills could be increased in proportion to the area,[115] but no mill could pump away the water if the wind failed and this was really at the bottom of the Fen problems. James Watt reckoned that a 10 h.p. steam engine would be needed to raise 500 cu. ft. of water 1.52 m (5 ft.) per minute, and it could work continuously.[116] William Gooch thought that a 20 h.p. steam engine was equal to a mill with a 12.19 m (40 ft.) sail in full velocity.[117] From this it would seem that the best fen windmills were producing about 40 h.p. when there was sufficient wind.

Compared with the Netherlands, English mills were at a disadvantage, for they usually pumped either into a river that was liable to flood or into a tidal reach where the level altered constantly. In either case, there were likely to be periods when the water outside reached a level higher than the windmill was able to lift it. In the Netherlands, most areas did not have the same difficulty with river floods; the tidal problem was overcome by building storage lakes into which windmills could pump all the time and from which the water was run into the sea through self-acting sluices at convenient stages of the tide cycle. A few storage lakes were constructed in England. The tide in the Wash might rise over 4.57 m (15 ft.) but a windmill with a scoopwheel could raise water only about 1.52 m (5 ft.). When the tide was favourable, there might not be enough wind, or the wind might come when the tide was up and in this case 'the engines must be gentle spectators till the Sea gives way'.[118] High floods in the rivers prevented windmills working, even with a good wind, and sometimes stopped steam engines, as happened to the Stretham engine in 1852.[119] One important reason for improving the River Ouse with the Eau Brink Cut in 1820 was that floods would escape more quickly, lowering the 'head'.

Subsidence of the general land level through better drainage, together with reduction of the peat, aggravated the situation. This could be countered by 'double lift' mills on the Dutch principle, although ring canals do not seem to have been used. The intermediate storage basin between mills could be from 100 m (100 yards) long in the case of the Waterbeach Level to nearly 900 m (½ mile) at Nordelph, and about 10 to 20 m (10 to 20 yards) wide, lined with clay, to prevent water soaking back through the peat. The English had known about this system from at least 1665,[120] but it had to be applied with increasing frequency around 1800. The Littleport and Downham District built a double lift mill in 1812 to work with their old Wey Next mill,[121] and the Waterbeach Level installed an 'inner mill' to lift water to their existing Dollard mill in 1814.[122] The little Redmore district of 1,780 acres was drained by a pair of double lift mills until 1878, when they were replaced by a steam engine and centrifugal pump.[123] Sometimes there was a double lift system in disguise. In the Waldersea and the Deeping Fen districts, individual farmers had their own small mills to lift water from their lands into the main drains while the drainage authorities maintained larger ones to pump the water into the river. In Deeping Fen, some smaller mills remained long after steam engines had been introduced.

Inadequacy of the mills stimulated people to search for improvements. One was Eckhardt's tilted scoopwheel. He took out a patent in England, where one mill may have been built in Deeping Fen and certainly one near Wisbech[124] and another on the Witham erected by Mr Chaplin of Blankney. On his mill, the sails went 'seventy rounds' and the wheel raised 60 tons of water 1.22 m (4 ft.) every minute. It drained 1,900 acres and probably cost £1,000.[125] Accounts of the performance of these wheels varied: 'Mr Weston who built it [the one at Wisbech], assured me that it is unquestionably superior to the vertical; but the millwrights of the country, who cannot execute them well, are averse to the use, and will not let them have fair play if concerned'.[126] At Deeping Fen, an Archimedean screw was tried in 1741 but subsequently abandoned.[127] A small drainage mill, 2.44 m (8 ft.) across at the base and 7.62 m (25 ft.) high to the top of a sail, drove one at East Tilbury on the Thames.[128] With Archimedean screws, it was more difficult to vary the height to which the water was delivered and this may have been one reason why they were never popular in England.

Time and time again there were tales of woe about the condition of land kept in little better than a half-drained state. Too often, when rain came there was no wind, or a period of calm followed an abundant fall.[129] Mr Scott of Chatteris had known mills to stand idle for two months when most wanted.[130] The situation was often worse after prolonged frost and snow for when the thaw came the mills had been unable to work, being frozen up, and could not lower the water level. The snow melted and ran off into drains which were already full and so flooded the land, while the rivers were in spate with floods from the high lands and were too high for the mills to raise their water. In such conditions, an inundation frequently took place.[131]

A correspondent of Horace Walpole, Mr Cole, moved to Waterbeach in 1767 and his experiences may have been typical. He wrote in June 1769,

> Since I wrote the former part . . . it was most seasonable to revive my spirits, at that time much depressed by fears and apprehensions of inundations over my estate, great part of which has been drowned these two years and is now getting dry again . . .
>
> However, I hope I am safe, bating the fright. This parish I now inhabit was not so lucky. On Monday night the bank of the river blew up, and has overflown a vast tract of country in this neighbourhood. I was all day on the water to see their operations; but they will hardly be able to stop it in three or four days. The mischief was occasioned by the rain on Sunday last. Longer rains I often remember, but never any so violent for the time it lasted, which was from Saturday night at 10 o'clock till Sunday afternoon about three or four. All this part of the country is now covered with water, and the poor people of this parish utterly ruined. I am determined to sell my estate in this country. Every shower puts me on the rack, and I have suffered exceedingly for these last four or five years, besides the continual uneasiness it occasions.[132]

A little over a year later, in November 1770, Cole wrote that, for the third time in six years, his estate had been drowned and this time worse than ever. In the following April, he wrote that he did not know when the country would be dry again as a result of those floods in November.

> This has been a constant damp upon my mind, and every shower put me on the wrack. Thank God I have got rid of this plague and anxiety by parting with my estate, which instead of being service, was a continual uneasiness to me – and of no great advantage. Within these three months, in consequence of these calamities, one tenant broke, by whom I loose about £400.[133]

Money was spent on drainage schemes that looked well when written as Acts of Parliament but were 'vain and nugatory'.[134] The burden of taxes was immense; in fact 'a Windmill drainage is the most imperfect of all modes and in many cases the adoption of such a mode may be said to be a useless waste of money'.[135] It was only the natural fertility of the land, which in one good year would recompense the fen farmer for the care and hazard of many bad ones, that kept him fighting against the floods.[136]

One estimate of the number of windpowered drainage mills in the Fens gave 700.[137] In 1748 the Rev. Thomas Neale, rector of Manea, said there were no less than 250 in the Middle Level and 50 in Whittlesey, many of which had been erected within the previous thirty years. When he was riding from Ramsey to Holme, he counted forty. 'There are between Ramsey and Old Bedford Bank, and upon the Forty feet, Sixteen feet and Twenty feet [all drains], and to Salter's Load in Well Parish fifty-seven'.[138] The Littleport and Downham District had seventy-five,[139] and those districts draining by the Nene about sixty.[140] One figure for Deeping Fen gave fifty in 1815[141] and another forty-four which developed 400 h.p. but none remained around 1848.[142] There were still some mills around 1878 on the Forty Foot drain which did their work well 'when the wind blew'.

> There is something touching about these old fellows, despite their occasional uselessness. In these days of rapid change, they still serve to remind one of the days of the past, and, like musty old tombstones in a country churchyard, recall to mind the memories of bygone worthies.[143]

In 1888, the Middle Level had some 40 windmills left at work, many with pulleys so that they could be driven by steam engines when the wind failed.[144] But by this time, most districts had replaced their mills by steam pumping stations. The building of Soham Mere mill in 1867 at a cost of £1,000 was an anachronism and the millwrights lost money on it. In the 1920s, all that remained at work were the two large double lift mills at Nordelph, the one at Soham Mere and a few small field mills in places like Adventurers' and Wicken fens.[145]

Most of the early steam engine pioneers thought that their machines might pump water off flooded land.[146] Thomas Savery envisaged draining 'Fens and *Marshes, etc*' with the one he patented in 1698, and in 1734 Marten Triewald suggested that Thomas Newcomen's engine could drain boggy and marshy countries. James Watt received an inquiry for a drainage engine in 1782 and wrote to Matthew Boulton that he considered the 'fens to be the only trump card we have left in our hand'. Further inquiries in 1789 from the Commissioners of Middle Fen near Ely and in 1790 from Marshland Fen, Norfolk, came to nothing partly because Boulton & Watt found they were expected to help finance the projects.

Writers about improvements to fen agriculture, such as William Gooch and Arthur Young, advocated steam engines over the next few years. One reason was that agricultural yields could be dramatically increased by mixing the peat with clay to neutralise the acid and produce a good tilth where excellent wheat could be grown. The clay was either dug up from beneath the peat or carted onto the land but the expense of doing this was only worthwhile if the drainage could be guaranteed. Should the acid 'soak' rise up and flood the land, the improvements would be destroyed. Windpower was too unreliable and steam engines were seen to be the answer. When John Rennie was asked to draw up schemes for improved drainage in areas like the Waterbeach Level, Borough and Deeping Fens, he recommended that steam engines should replace windmills. To the Commissioners of Borough Fen, he pointed out that a steam engine could work during periods of enforced idleness for the windmills and so some water would be cleared away, enabling the mills to perform better when there was a wind. One large steam engine could replace many mills with consequent savings in men's wages. Although there were the costs of the fuel to be considered, steam engines could guarantee the drainage, something the windmills had failed to do.

A steam engine, installed at Hatfield Chase in 1813, was probably the first one specifically for land drainage in England. The first one discovered so far in the Great Level of the Fens was at Sutton St Edmund soon after 1816. The Ten Mile Bank engine for the Littleport and Downham District was completed by March 1820 and a little later in the same year one was finished for Borough Fen close to Sutton St Edmund. John Rennie's advice was taken by the Swaffham and Bottisham district, where an engine was built in 1821. Rennie's death in 1821 prevented him supervising the erection of two engines, one of 60 and the other of 80 h.p., at Pode Hole in Deeping Fen which were completed in 1825. They were not very successful at first because the drains conveying the water to them were too narrow. Trials were carried out in July 1830 after which the drains were widened and the engines then worked better.

While two more small engines were built for March West Fen in 1826 and Pinchbeck South Fen in 1830, it was the 80 h.p. Hundred Foot engine for the Littleport and Downham Commissioners, finished in the spring of 1830, which really showed that steam power could effectively and economically drain the Fens. The original scoopwheel here was 10.67 m (35 ft.) in diameter and weighed 54 tons, much larger than in any windmill. At low tide, it could discharge 32,880 gallons of water a minute, an impressive performance. The engine house is still preserved. In 1831 the tax was removed on coal imported by sea; this encouraged the erection of more steam engines in the next few years, including the Stretham engine which has been preserved with boilers, engine and scoopwheel. One by one the different districts switched to steam because the guaranteed drainage enabled the black fen farmers to clay their land and produce rich crops of wheat. Through the national system of canals and later the railways, this grain could be transported to the expanding industrial cities of the midlands and northern England, so the Fens saw a period of great prosperity.

The Norfolk Broads

In Norfolk, the rivers Ant and Bure combine to form the Thurne which meets the sea at Great Yarmouth. Just below Norwich is the confluence of the rivers Yare and Wensum which, together with the River Chet, also make their way to the sea at Great Yarmouth through Breydon Water. The River Waveney joins them in Breydon Water although it must have had its exit through Oulton and Lowestoft at one time. They all meander through marshlands formed by silt deposits on which freshwater bogs grew up and created peat beds. Parts of the peat were totally excavated towards the end of the Middle Ages, presumably for fuel, so that large areas of open water have been left, forming the famous Broads.[147] The method of peat extraction in the Fens was different, allowing peat to re-form so that lakes never developed. Generally the Norfolk river valleys are quite narrow and the marshlands on either side of their courses quite small. The higher land is excellent arable still noted for its grain crops. Until recently, the marshlands were kept as pasture, partly owing to the influence of London, where not only was meat much in demand but fodder in the form of hay and litter from sedge was needed for the capital's motive power for its transport – horses.

As long as grass remained the principal crop on the marshes, water levels in the drains did not need to be kept as low as in the Fens where arable farming became much more important. This meant that windmills were sufficient to drain the Norfolk Broads marshes for much longer than elsewhere. In addition, the winds here were stronger than in the Fens, taken over the year.

Wind speed	Holland	Fens	Norfolk Broads
6–8 m/sec.	1,332	1,778	1,226* hours
8–12	1,339	450	964
	2,671	2,228	2,190

* Figure for 7–8 m/sec. only.[148]

While the first figures for the Broads are for the 7–8 m/sec. (15–18 m.p.h.) band only, in the effective range of 8–12 m/sec. (18 to 27 m.p.h.), this area has twice as much wind as the Fens but still less than Holland. This, together with a low annual rainfall of about 63.5 cm (25 in.), goes far to explain why wind drainage mills continued to be built here until the opening years of the twentieth century. Because most of the marshy area lies quite close to the sea, the land might have been drained by natural means had not another factor, the tilting of the whole land mass of the British Isles, gradually made itself felt. While the west of Britain is slowly rising, eastern parts are sinking into the sea. This may account for the lack of any mention of windmills on the Broads until well into the eighteenth century. Also, the areas to be drained here remained small right up to the present time and individual farmers might not have had the capital to invest in such drainage equipment. High's Mill, on the Halvergate Marsh, drained only 200 acres and Flegburgh Mill 394 acres.[149]

There were also some larger areas and Boards of Commissioners similar to those in the Fens were set up in a few places.

The smock mill preserved at Herringfleet is of a very early type although it was built as late as 1830 (Figure 69). It has horizontal weatherboarding on the outside while inside there is a fireplace to keep the miller warm. The sails were never modernised and remain the common type covered with cloth. The cap is winded with a winch at the bottom of tailpoles in a manner similar to Dutch mills. Around the mill are posts driven into the ground to form anchor points. The common sails and the winch were the reasons for having a person to attend to this mill when it was working. A wooden housing at the side of the mill

Figure 69. Herringfleet Mill on the Waveney marshes is the sole survivor of a wooden smock mill with tailpoles, common sails and scoopwheel. The shed behind contained the modern engine and rotary pump (March 1990).

Figure 70. A case of 'adaptive reuse' for the tower mill at St Benet's Abbey was built around one of the arches in the gatehouse. It was used first as an oil and then a drainage mill (March 1990).

protects the 4.88 m (16 ft.) diameter scoopwheel. The few modern parts in the mill are the windshaft, the axle for the scoopwheel and the framing of the scoopwheel, all made of cast iron. The mill was replaced by a Ruston turbine pump in about 1950 after a working life of over 120 years. Today its white sails and black tarred body stand out attractively over the marshes.[150]

Rex Wailes asserted that 'it is safe to assume that the first mills were timber built smock mills'[151] but this might not necessarily be the case, for Norfolk had a long tradition of using bricks. Known dates for mills commence only in the middle of the eighteenth century. The one built into the gateway of St Benet's Abbey is the oldest with a date of 1735–40, but it may have been used first for crushing cole seed for oil (Figure 70). Wailes gives dates for Wiseman' Mill, at Ashby on the Bure, as 1753 and Brogave Mill, Walcot Marsh, near Sea Palling, as 1771.[152] All of these had brick towers and brick construction continued throughout the nineteenth century. Dates on other surviving towers range throughout the whole of the latter century and into the twentieth. Thurne Dyke Mill was built in 1820, Hunsett Mill at Stalham in 1860, Stracey Arms Mill in 1883. Martham has a plaque on the exterior, which reads 'Erected 1908, Saml Aldred Wm Bracey senr,' while another plate inside bears the legend '1912 by England of Ludham'. The mill at Horsey Mere was rebuilt in 1897 and

1912.[153] The last mill to work under wind, Ashtree Farm near Great Yarmouth, was built in 1912 by Smithdales, the millwrights of Acle,[154] showing wind-powered drainage mills were still a viable proposition. Faden's Map of Norfolk, published in 1797, has forty-six drainage mills in the east of the county and it is thought that most in existence then are marked. The first edition of the Ordnance Survey maps of Norfolk, published in the middle of the nineteenth century, shows ninety-two drainage mills and Wailes in his survey during the early 1950s reached a total of 107. Not all of these mills were the same in each survey, so there may have been nearly 140 in all.[155]

High's Mill, on the Halvergate Marshes, bears evidence that at one time these early mills must have had tail poles, but all that remains here is the long crossbeam behind the front bearing of the windshaft protruding from either side of the cap where one set of braces was attached. All the other mills that survive today, in various stages of preservation, were fitted with fantails, except this and Herringfleet. Again, all the early mills must have had common sails but, once more, old photographs of mills and existing remains show that most were modernised with patent sails. This meant that the mills could be left to work more or less on their own without constant supervision. Unlike earlier mills in the Fens, no millers lived in the majority of Norfolk mills. However, Eastfield Mill at Hickling was taller than most, with four floors where a Mr and Mrs Gibbs brought up their family of twenty-one children.[156] Conditions inside must have resembled those of the old lady in the nursery rhyme who lived in a shoe.

Like Herringfleet, the early mills must have driven scoopwheels. That at Ashtree Farm was 4.57 m (15 ft.) in diameter, 17.8 cm (7 in.) wide and with 15 r.p.m. of the sails lifted 8 tons of water a minute. This was probably an average size. Cadge's Mill on Reedham Marshes had a wheel 5.49 m (18 ft.) in diameter and the mill on the Limpenhoe Marshes one of 5.79 m (19 ft.) but these were exceeded by the exceptional one at Berney Arms (Figure 71), 7.32 m (24 ft.) in diameter.[157] Scoopwheel widths varied from about 12.7 cm (5 in.) to 27.9 cm (11 in.) at Berney Arms and 45.72 cm (18 in.) at Mutton's Mill, Halvergate Marsh. Surviving wheels have cast iron frames. Smaller ones had a single frame and a single start post to which the ladle boards were nailed. One 5.79 m (19 ft.) diameter wheel can be seen at Lockgate Mill on Breydon North Wall. The Five Mile House Mill was larger and had a pair of frames and twin start posts, which was probably a more common arrangement. While most wheels were situated alongside their mills in brick troughs with wooden covers, a few were inside. Apparently these tended to ice up more quickly in winter because the draught was funnelled through them, so bags of salt were kept in these mills. At Mutton's Mill, the wheel turned at the same speed as the sails but general practice made the wheels rotate more slowly.

In contrast, the centrifugal pump, which had to rotate quickly, began to supersede the scoopwheel in the 1850s. The decision was taken in 1851 to drain

Figure 71. The first machinery at High mill, Berney Arms, was edge runners for grinding cement clinker and the drive was later extended to the scoopwheel on the left. The mill is fitted with full patent sails and a fantail so it can be left running unattended (March 1990).

Whittlesey Mere, in the Fens, but a natural gravity drainage proved to be impossible. In 1852, Easton & Amos installed a 25 h.p. steam engine driving one of Appold's newly invented centrifugal pumps which proved capable of discharging 16,000 gallons an hour against a 1.83 m (6 ft.) lift.[158] This type of pump had a vertical spindle driving a rotor at the bottom of a well either of masonry or a tube of metal (Figure 72). Early pumps had twin inlets, one above and one below the rotor, but later ones had only the lower. Water was centrifuged up the well and passed out through a sluice door similar to those in front of scoopwheels, which shut when the pump was not running. The rotor was placed below the level of the water in the drain so it always lifted the same volume.[159]

There was great controversy over these turbine pumps. One advantage was that both the housing and the machinery were much lighter than scoopwheels and so the original capital cost was less. Because there was less mass to start rotating, the pumps began to operate in lighter winds. When the level in the drain fell, the pumps always lifted the same volume of water because the inlet was below water level while the ladles on scoopwheels dipped less. Another advantage was that, when the peat shrank and the level of the land became lower, the pumps could be lowered much more simply and cheaply than scoopwheels. The one at Whittlesey Mere proved this because it had to be lowered

Figure 72. The gearing in the top of the casing housing the centrifugal pump at Thurne mill (March 1990).

twice within the first 25 years of its life. At first the lift was 1.22 to 1.52 m (4 to 5 ft.) but by 1877 it was over 2.74 m (9 ft.).[160] Scoopwheels were claimed to be more efficient in removing the highest water when the drains were full. Relative sizes of a scoopwheel and a pump to be driven by a 14 h.p. steam engine were:

	Centrifugal pump	Scoopwheel
Power of engine	14 h.p.	14 h.p.
Lift	11 ft.	11 ft.
Diameter of wheel	3′ 4″	40′
Width of wheel	8¼″	18″
Number of revolutions	180	4
Diameter of outlet pipe	2′	2′[161]

It was generally agreed that the pumps would work in lighter winds when scoopwheels remained inoperative and that in steady winds they would throw out twice as much water.[162] However, they did not work as well in gusty winds, when presumably the scoopwheel ran more steadily through its greater weight. The preserved mills at Stracey Arms and Thurne Dyke both have centrifugal pumps. The mill at Thurne Dyke shows another feature, that of being heightened with a parallel brickwork section above the original tower to take larger sails, possibly when the pump was installed.

As in the Fens and in the Netherlands, steam began to replace windpower, but never to the same extent. High's Mill, Potter Heigham, drained over 1,000 acres with a turbine pump and was assisted with a steam engine, while on the Reedham Marshes a steam plant was built beside Polkey's Mill. Haddiscoe Marshes were drained by a complete steam pumping unit but then the small compact heavy oil engine appeared. Sometimes an engine would drive an existing pump at a windmill but often a more modern pump would be installed in a separate building. An early oil engine unit survives on the Reedham Marshes which in its turn has been replaced by electric pumps placed in one of the old windmill towers. Although one electric pump was installed around 1932,[163] the spread of the grid after the Second World War hastened their introduction and so brought the demise of the remaining windmills. The windmill on Mautby Marsh has been replaced by electrically driven, modern steel Archimedean screws. On 1 February 1953, the sails and part of the windshaft were blown off Ashtree Farm Mill, the last windmill to remain at work in the Norfolk Broads, so ending the era of windpower.[164] The preserved mills cannot do any useful work, for their pumps have been blocked off or disconnected; but they remain to remind us of a bygone age.

CHAPTER 7

Windpower for industry

Sawing

The low-lying promontory of the Zaan situated between the North Sea and the IJsselmeer was well placed for wind and this must have been one reason why the industrial windmill evolved here around 1600, launching a form of Industrial Revolution. Between 1500 and 1650, land reclamation added about 40 per cent to the area under cultivation but this was won only through large-scale investment in drainage works, which had to be maintained through taxation. The population increased by something like 150 per cent. Some people were employed in the fishing industry and the shipping trade for North Holland became the major recruitment area for seamen and fishermen. The industries were linked together through wood: wood for houses, wood for sluices and locks in the drainage works, wood for the 2,000 ships built annually and wood for windmills. The wind-driven sawmill became one vital link in this expansion while the other industrial mills were closely linked to agricultural products. Oil mills crushed seeds from flax, hemp mills prepared hemp fibres, papermills had as their raw materials ropes and sails from ships, and corn and barley hulling mills fit into this pattern too. The Dutch world-wide trade was the basis for dye, tobacco, snuff and cocoa mills but their numbers were fewer.

The Dutch industrial windmill was one of the great achievements of traditional millwrighting in wood. The technology was closely linked with that of waterpower but had to be modified to suit the particular restraints of a cramped building that needed to present as small as possible an obstruction to the wind. The framing for the body of the mill itself was similar to that of smock mills (see Figure 82 below) and the machinery was adapted from a few basic concepts. One was the use of the crank for driving reciprocating saws and the Dutch sawmill will be examined first. As the Dutch had no supplies of good timber in this region, it all had to be imported. The Baltic was one source but most arrived in rafts which were floated down the Rhine and the wood was left in the water before it was sawn. Sawmills were always situated along a waterfront, sometimes on islands so that the logs could be drawn up into a windmill whichever direction it was facing.

In hand sawing, the log was placed over a pit. One man stood on top and pulled the saw up while his less fortunate partner – the pitman – stood below and pulled the saw down again, with sawdust falling over him. As the cut progressed along the log, so the men moved across the pit. In the first mechanical sawmills, reciprocating saws were still used and there had to be a pit beneath the main sawing platform because the saws moved vertically; but the blades went up and down in the same position so the log, mounted on a movable frame or trolley, was drawn past as the cut advanced. This meant that the sawing platform had to be able to accommodate first the length of the unsawn log, then the sawing area, and then the finished planks.

The oldest known waterpowered saw was recorded in Normandy in 1204 and another was illustrated in the sketchbook of Villard de Honnecourt around 1270. By the late fifteenth century the basic patterns had been established. The frame holding the blade might be drawn down by cams projecting from the axle of a waterwheel and was returned by a flexible spring pole above, or it might be raised by cams and allowed to fall and make the cut by its own weight. In Francesco di Giorgio's treatise on machines and in some of Leonardo da Vinci's sketches, there are illustrations of waterpowered machines with cranks connected to rods which moved rectangular frames between vertical guides. In each case, one blade was set in the frame and the wood to be cut placed on a trolley moved forward by a ratchet. Later a gang saw with a number of blades fixed side by side was introduced for sawing planks more quickly.[1]

The first windpowered sawmill was invented by Cornelis Corneliszoon of Uitgeest (Figure 73). He obtained a patent, to last for twelve years, on 15 December 1593 from the States of Holland for a windmill to saw all types of wood. A sketch shows a hollow post mill like a wipmolen driving a horizontal crankshaft which worked the vertically reciprocating saw blade. On the shaft was also a gearwheel, later replaced by a ratchet, which connected through reduction gearing to a rack for pulling along the wheeled trolley on which the timber was secured. There is no building surrounding the saw bench. This mill, 'Het Juffertje', The Damsel,[2] was built in 1594 or 1595 and in 1596 was moved to Zaandam. His next design in 1597 had a triple-throw crankshaft and all the sawing apparatus was brought inside the mill. He constructed another mill which was probably quite small because, while the normal tax to operate a windmill was £50 per annum, Cornelis, as a millwright and not a mill owner, paid only £2. It was transported to the village of Wormeer on a raft in 1601. Cornelis took out a further patent at Utrecht in 1600 and it was after this that the '*paltrok*' was developed. Dirk Sybrandts was associated with making several improvements, one of which included increasing the number of saws.[3] The name

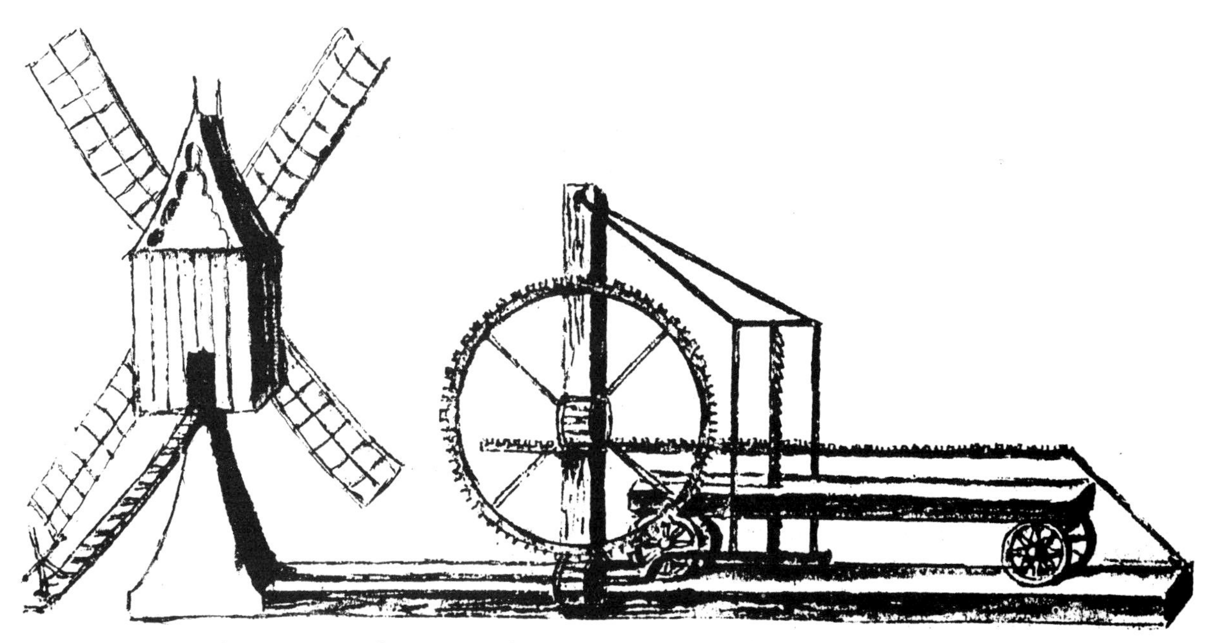

Figure 73. The drawing in the first patent of Cornelis Corneliszoon for a sawmill (1593).

of the paltrok is said to have been derived from the flared coats, the *Palts-rokken*, worn by Mennonites who had emigrated to the classical country of liberty and had settled in the Zaan area.[4]

The paltrok was based on a modified form of post mill (Figure 74). Because the logs needed to be sawn lengthways into planks, the sawing platform was placed at right angles to the windshaft with the sawing mechanism situated in the middle under the main body of the mill. Wings with roofs and fronts were added on either side to give protection from wind or rain, hence the allusion to Mennonite skirts. The sawing platform was a couple of metres above ground level (7 ft.) and was left open at the back. The entire mill had to be rotated to face the wind and pivoted on a central bearing. The main post supported the normal crown tree above the level of the sawing platform. To help balance the mill, there might be two sets of sawing frames, one either side of the post, or sometimes three: a large set to one side and two smaller ones on the other. The mill was also mounted on a large ring of rollers on a brick base below the sawing platform, which gave great stability. A door was fitted in the front panelling to give access to a staging for adjusting the cloth on the sails. At the back, the winch for winding the mill was fitted onto the framing carrying the rollers so that long tail poles were not needed. The sails turned the windshaft in the usual way but the brakewheel engaged a second vertical wheel which was keyed onto a horizontal crankshaft just below and parallel to the windshaft. Vertical connecting rods operated the sawing frames. A crane, with a winch driven through a ratchet, hauled a log onto the sawing platform and placed it on the trolley which was moved by another ratchet. The cut was taken on the downstroke of the frame and blades through their weight, while the wood was moved forward on the return. Saw-blades were spaced out across the vertical frames at distances needed for the width of the planks and held in position by wedges (Figure 75).

Although they seem massive machines, paltroks were used for sawing the lighter wainscot or planks for furniture or panelling. In 1604, the first smock type of sawmill, the Grauwe Beer, was built and was soon followed by others (Figure 76). With their larger span of sails and sturdier construction, they could tackle bigger logs. In these mills, since only the cap had to be turned to face

Figure 74. The paltrok, Poelenburg, now preserved at the Zaandse Schans museum. The side extensions to protect the sawing platform give it the distinctive shape (March 1992).

Figure 75. The three sawing frames in the Poelenburg mill. The two on the right are fitted with sets of blades. The mill was running when this photo was taken (March 1992).

the wind, the sawing platform could be built inside a long shed. Logs were drawn out of the canal or storage pond up a ramp at one end and passed to the sawing frames in the middle. The layout of the machinery was basically the same as in the paltrok but the crankshaft was driven by a short vertical shaft from the windshaft in the cap. While this did mean that there was loss due to friction in two more gears and bearings, the working space in the mill could be much larger. Usually the smock sawmill was built in the form of a hexagon so that there were fewer main posts to obstruct the working area. With these new mills, production rose by 3,000 per cent. With hand sawing, 60 beams or trunks would take 120 working days compared with 4 to 5 using windpower. By 1630, there were 83 sawmills to the north of Amsterdam of which 53 were in the Zaanstreek. The peak was reached in 1731 when there were 450 in the Netherlands of which 256 were in the Zaanstreek, but few remain today.[5] The paltrok 'Poelenburg' now preserved at Zaandse Schans was built in 1867 and moved to its present site in 1963. There is another in the Open Air Museum at Arnhem. Two smock mills in Leiden, 'De Herder' built in 1884 and 'd 'Heesterboom' built in 1804, have been preserved as well.

Windpowered sawmills were rare in Britain. One was erected by a Dutchman near London in 1663. The supply of great beams for shipbuilding at this period was barely sufficient to keep up with the demands of shipwrights and this mill caused a sensation. However, the hostility of the hand-sawyers was so great that they attacked it with axes. The Dutch millwrights left the scene hurriedly and the mill stood abandoned for several years.[6] In 1687, the Duke of Albemarle, recently appointed governor of Jamaica, was granted a patent for the right of 'erecting saw mills to move by wind or water' in the American colonies and plantations with the exclusion of New England.[7] It was not until 1767 that a

Figure 76. The later smock sawmill, De Herder, at Leiden. The sawing frames are housed below the smock mill while the long shed contains the trolleys (May 1983).

London timber merchant, John Houghton, had the support of the Society of Arts and erected a sawmill at Limehouse, London. The work was directed by James Stanfield who had gone over to Holland and Norway where he learnt how to construct and manage such a mill. It was no sooner finished than a mob assembled and burnt it to the ground. The principal rioters were punished and a new mill was built, at the nation's expense, which worked without further molestation.[8] An 1819 advertisement for an eight-sailed sawmill and glue factory in Hull described it as 'a most complete piece of mechanism and has been used principally for cutting Veneers for the Leeds and Norwich Bombazeen manufacturers, it also works a bone mill and pumps'.[9]

Scotland's close links with the Continent in the seventeenth century helped to introduce new technology. A charter of 1638 reserved the liberty to build either a water- or a windpowered sawmill on Deeside, Aberdeenshire.[10] Even if a windmill were not erected here, there were certainly windpowered sawmills at both Leith and Airth before the Act of Union in 1707.[11] In the 1790s there was a windmill at Garmouth containing thirty-six to forty saw-blades. In the early years of the nineteenth century, there were mills at Ceres (Fife) and Peterhead, while a wood-turning windmill at Crieff was demolished about 1854.[12]

By this time circular saws had been invented. Marc Isambard Brunel, who designed the production line system for making ships' blocks at Portsmouth, took out a series of patents between 1805 and 1813 which included circular saws and their use spread rapidly.[13] In the later years of the nineteenth and early twentieth centuries, a few corn mills were converted to saw wood, probably with circular saws. A list of industrial windmills in Norfolk includes a dozen used for sawing which all date after 1840.[14] There were a couple of six-sailed sawmills in Kent. That at Bethersden had originally been a corn mill before being moved and converted into a sawmill before demolition in 1896. The one at Great Chart was erected by a builder for the purpose of sawing wood. It is known to have been working in 1888 and remained out of use for many years until being pulled down in 1928.[15] With a circular saw bench, it was quite easy to take a belt drive to the sawing spindle from some convenient piece of rotating shafting as, for example, at Punnett's Town mill, Sussex.[16] Yet it cannot be said that windmills were used extensively for sawing in Britain. Some Finnish windmills served the dual purpose of wood sawing and corn grinding but these were nowhere on the scale of Dutch industrial sawmills.[17]

Crushing and pulping

The millstone, set on edge in the vertical position and pushed round in a circle to crush material by its weight, was an early invention. These 'edge runners' or 'kollergangs' had a pair of stones by the seventeenth century, possibly patented by Cornelis Corneliszoon in 1597,[1] which could be connected to any convenient form of rotary power. They might be used in conjunction with stamping mills for crushing and pulping. Vertical stampers may have been derived from the pestle and mortar and there were stampers with horizontal shafts which could have originated with the hand-held hammer. Both types were mechanised through camshafts turned by power. The Dutch industrial windmill took over this earlier technology.

Dyestuff mills

Oil and dyestuff mills employed vertical stampers and kollergangs. In dyestuff mills, the raw material, wood, first had to be broken up into small pieces. In the mill, 'De Kat' at Zaandijk, blocks of logwood were cut up by hand and transferred to a '*kapperij*'. This was a Dutch invention of the later part of the seventeenth century, used also in papermaking. A camshaft high up in the mill operated four vertical stampers. Their feet had cutting knives which chopped up the wood thrown into a tub. A ratchet device worked by another cam and vertical lever slowly rotated the tub. When the wood was reduced to shavings, it was taken to the kollergang. Gearing from the main shaft drove the edge runners, which crushed the wood to powder by their weight. One stone was set further out from the centre than the other so that a wider area was covered. Scrapers helped to push the material back under the stones. The resulting powder was ready to be used for dyeing. In the Noorderkwartier of the Netherlands, there were twenty-one dyestuff mills in 1731.

As Josiah Wedgwood expanded his pottery business, he found that the scale of manufacture needed power-driven machines, both for flint grinding and also to prepare the colours. Erasmus Darwin proposed his novel form of horizontal windmill and in a letter to Bentley in March 1768, Wedgwood mentioned this mill as a possible way for grinding colours, but there the matter was left for over ten years. Darwin did not abandon his project and continued to conduct experiments. On 20 August 1779, Richard Lovell Edgworth wrote to Wedgwood:

> As Journeyman to Dr Darwin I have tried a very great number of Experiments to determine whether our horizontal Mill might be made to advantage for your purposes: And by several ingenious contrivances which the Dr Proposed and which I subjected to the test of Experience this species of Machinery may be made to exceed any other horizontal wind-mill in the proportion of four to one or perhaps in a yet higher ratio.[2]

James Watt was also drawn in to help and Wedgwood was persuaded to adopt this type of mill for grinding his flints and colours. Later it was replaced by

Figure 77. The kollergang for crushing the seeds in the oil mill, De Bonte Hen, Zaandse Schans museum (March 1992).

one of Watt's rotative steam engines. The first steam-driven dyestuff mill in the Zaan area was 'De Blauwe Hengst', the Blue Stallion, in 1835, but the owner soon changed back to windpower.

Oil mills

Oil was extracted from the seeds of plants like colza, flax and rape and was used for cooking and lighting. First the seeds were squashed under edge runners. The resulting pulp was placed in a pan set over a fire and heated, being stirred all the time, then taken to stampers for pressing in a complex system (Figures 77–80). Large baulks of timber formed the bottom troughs of stampers which were carved out into suitable shapes for the two stages in this part of the process. In the preliminary pressing stage, the pulp was placed in a woollen bag, packed in a kind of matting made from horsehair and then wrapped in a piece of leather. This package was put into a rectangular hole carved in one baulk together with wedge-shaped blocks of wood. Often two packages were pressed together, one at either end of this hole. One stamper acted as a ram and hit the larger end of one of the wedges to squeeze and press the oil out of the packages. This stamper needed to act fairly slowly so it normally had only two cams to operate it instead of the more usual three. A dial mechanism counted the number of strokes, usually fifty, and rang a bell on completion. The miller stopped that stamper and engaged the second which knocked down with a few strokes a second wedge with reverse taper, thus freeing the wedges and packages so they could be taken out. The oil ran away through holes in the bottom of the baulk of timber into

Figure 78. At the back, the lower part of the vertical shaft and gearing to drive the kollergang and in the foreground the camshaft for the stampers (March 1992).

Figure 79. The pulped seeds are wrapped in special bags and put at either end of the pressing block. The tapered wedge is driven down by one stamper to press out the oil and then the wedge with reverse taper next to it is knocked down by another stamper to free the packages. The ear protectors hanging on the framing are used today to prevent deafness (March 1992).

pans beneath.[3] To obtain the maximum amount of oil, the cake of pressed seeds was taken out of the package and put into circular pot holes hollowed out of the timber baulk in the second set of stampers. Here a single stamper to each pot broke up the cake into a powder. This was heated once more and placed in another package for a second pressing. Afterwards the cake would be taken out and sold as cattle feed. The noise of these mills at work with their two different sets of stampers was terrific and could cause deafness.

This system of oil extraction probably did not originate in windmills and may not have been devised in Holland either. In the north of France, there were post mills which contained pressing mechanisms. Presumably the first stage with kollergangs was done elsewhere because the space available inside a post mill was too restricted and the weight of the large stones would have been excessive. The post mill, Moulin de Bourbourg, shows the ingenuity of the millwright. The windshaft was used as the camshaft for the second set of stampers and their breaking pots. As of course the windshaft was inclined, the length of the vertical hammer shafts had to be increased from the back towards the front of the mill. Across the front of the mill was a pair of pressing blocks with their camshaft across the mill below the brakewheel driven by a short vertical connecting shaft. A drive from this gearwheel turned the stirrer in the heating pan and, in this particular mill, also a modern roller seed crusher.[4] Rex Wailes mentioned two more of these oil post mills, which had burnt down prior to his visit in about 1937, so the type cannot have been that rare.

Figure 80. A second set of stampers break up the pressed cake in pot holes for reprocessing (March 1992).

These mills could pre-date the larger Dutch ones, for Lief Jansz Andries van Moerbeeck came from Flanders and erected the first oil mill in the north of Holland at Alkmaar in 1582. Tradition states that this was a type of wipmolen.[5] Cornelis Corneliszoon took out a patent on 6 December 1597 for an oil mill with edge runners and a single vertical stamper in a tower or smock mill. In large industrial smock mills, the main body was built above the working area which extended on either side. All the equipment was at ground level with the kollergang which was driven by reduction gearing from the main vertical shaft. In some mills, the drive to the kollergang could be disengaged so that the mill could be started slowly in light winds and the kollergang engaged when the mill was running. High up in the mill was another gearwheel which meshed with one on a horizontal camshaft for operating the stampers. In larger mills, there might be two sets of presses and six stamps and the camshaft would stretch across the whole body of the mill.[6] These mills are fine examples of the millwright's art with their massive framing both for the mill itself and for all the machinery. In 1731, there were about 140 oil mills in the Noorderkwartier which remained competitive for well over a century, for even in 1857, the cost per ton of oil produced in a windmill was less than in a steam-driven mill.

Standing on a hill overlooking part of the harbour at Stockholm is an eighteenth century oil mill covered with copper sheathing, similar to those in the Netherlands but with an extra drive to a set of reciprocating saw-blades placed close together as a form of rasp to break up logwood for dyes. In England, there were some wind powered oil mills. The Dutch are said to have built the South Lynn oil mill in 1638 as a smock mill and Millfleet Mill was recorded at King's Lynn in 1725.[7] Samuel Hartlib, writing about husbandry in Brabant and Flanders in 1650, recommended wind and watermills for turning flax seed into oil.[8] On the Norfolk Broads, the windmill built between 1735 and 1740 into the gatehouse of St Benet's Abbey may have been used first as an oil mill for colza oil although, when the sails were blown off in a gale in 1863, it appears to have been a drainage mill with a scoopwheel.[9] Another mill is recorded near Newcastle upon Tyne for oilseed crushing and oilcake making.[10]

John Smeaton built a windmill at Wakefield in 1755, which could have been his first. This is described as an oil and logwood mill for a Mr Roodhouse. In this instance, the cap was turned by gearing driven by an endless chain descending to a balcony.[11] Some of Smeaton's drawings survive and must be the earliest scale working drawings of an English windmill. They show that some of the millwork was made from cast iron and Smeaton claimed that he was the first person to use it. The apparatus for obtaining the oil consisted of a kollergang and vertical stamps. The sails were built according to the shape he had found most efficient in his experimental test rig and had a leading side wider at the tip than at the centre, a practice he does not seem to have followed later.[12] While he built several other oil mills driven by either water or steam, his next oil windmill was Sykefield Mill at Austhorpe, Yorkshire for John Brooke, a merchant in linseed oil, who had married Smeaton's daughter Ann in 1780.[13] The mill was completed in 1781 and was one of the first to be equipped with five sails. The windshaft was cast iron and had the cross to which the sail backs were bolted cast integral with the shaft as was the boss for mounting the brakewheel.[14]

Oil pressing equipment does not seem to have been altered throughout the eighteenth century and into the nineteenth. Jamieson's *Dictionary of Mechanical Science*, published in 1827, has an illustration of water powered stampers and presses similar to those in Dutch windmills. There were twenty waterpowered oil mills in Scotland but no windmills.[15] The steam engine ordered in 1783 by Cotes and Jarratt for their oil mill at Hull was the first in a new line of rotative engines by Boulton & Watt, for it had a double-acting cylinder and should have been fitted with a rack and sector linkage between the piston rod and the beam but was actually installed with Watt's newly patented straight-line or parallel motion. It was the first of the standard type of steam engine which would drive so many textile and other mills in the first half of the nineteenth century.[16]

Flint-grinding mills

The use of powdered calcined flints in pottery can be traced to John Dwight of Fulham, who wrote in his notebook in 1698 when giving the ingredients for a white body, 'Calcined beaten and sifted flints will doe instead of white sand and rather whiter but ye charge and trouble is more'.[17] The flints had to be roasted in a kiln to calcine them and then reduced to a powder before being mixed with water in the clay. Flints were being used in the Staffordshire area by 1720 when pestles and mortars for preparing them were listed.[18] This method was slow but, by 1725, waterpowered mills were in operation using vertical stampers. The resulting dust proved to be a great health hazard and in November 1726 Thomas Benson, who described himself as Engineer of Newcastle-under-Line [*sic*] in Staffordshire, patented flint-grinding in water

> The method hitherto used . . . by pounding or breaking it dry, and afterward sifting it through fine lawns, which has proved very destructive to mankind, insomuch that any person ever so healthful or strong working in that business cannot probably survive above two yeares, occasioned by the dust sucked into his body by the air he breathes, which being of a ponderous nature, fixes there so closely that nothing can remove it, insomuch that it is now very difficult to find persons which will engage in the business.[19]

He proposed to crush the flints under iron edge runners and then put them in a large iron pan with water in which iron balls were driven round by arms attached to a central rotating shaft. The balls crushed the flints into a slurry 'as fine as oyl itself'. However, particles of iron from the balls caused discoloration in the pottery and, in a further patent in 1732, Benson proposed using granite balls on a similar bed.[20] Later the stone 'chert' was substituted because it was harder and purer than granite.

The second patent would have expired in 1746 and this must have been the type of flint-grinding mill installed by both Smeaton and James Brindley in windmills. The comparatively late introduction of this system of grinding must have meant that good waterpower sites were few and this may account for the use of windpower. Smeaton may have been connected with a flint-grinding windmill at Nine Elms shortly before 1754.[21] In 1758, Brindley erected a large windmill at the Jenkins near the present Burslem Town Hall for the potters John and Thomas Wedgwood. The vertical main shaft was extended to the base where it drove the arms in the pan to push round the chert blocks.

> On the first day the mill was set in motion, to the consternation of everybody and most so of its inventor, the sails were blown off during a high wind.[22]

The mill must have been repaired soon, for Brindley erected other flint-grinding mills driven by wind or water.[23]

Chalk and cement mills

The edge runner was also used for crushing chalk, both in its raw state to make whiting and when calcined to make cement. Edge runners over 1.52 m (5 ft.) in diameter with large iron bands encircling their circumference still lie abandoned outside the whiting mill at Hull close to the Humber Bridge (Figure 81). Cliff Mill, built in about 1810, is a tall brick structure and the holes in its side show where it was surrounded by a staging. It had five sails with roller reefing blinds but these were removed in 1925. Now both its cap and the low drying sheds have disappeared.[24] This was one of four such mills in Yorkshire, another being built at Beverley in 1837.

High Mill at Berney Arms, across the marshes from Great Yarmouth, has its own railway halt and is now in the care of English Heritage. On the far side of the river was a factory making cement from chalky mud dredged from the river. The cement clinker was ferried across the river to be ground at the mill under edge runners. The mill was built by Stolworthy's, the Yarmouth millwrights, with seven floors, but the year is not known, the only clue being a pencilled date of 1870 in the cap. The edge runners were placed on the second floor and there could have been other sets too, for this is a powerful mill with a sail span of about 21.33 m (70 ft.). Nowadays a fantail keeps the sails automatically facing the wind and all the sails are fitted with patent shutters so the mill can

Figure 81. On the outskirts of Hull stands the tower of Hessle chalk mill with edge runners abandoned outside (July 1990).

be left unattended to run day and night as a tourist attraction. The vertical iron shafting is now disconnected from the 7.32 m (24 ft.) diameter scoopwheel, a few yards outside the tower, which the mill drove after conversion to a drainage mill at some time.[25]

The Dutch had similar mills for grinding mortar and, in England, references have been found to windmills for making gunpowder, in Oxfordshire for grinding ochre and in other places for bark. There is mention of a starch mill at King's Lynn and bone-crushing mills at Norwich and Sculthorpe in Norfolk. More romantic perhaps were the tobacco mill at Devizes and the snuff mill at Cotham, but numbers of such British industrial mills always remained few. In the Netherlands, the only industrial windmills to increase in numbers between 1731 and 1795 were snuff and tobacco mills which rose from thirty to probably thirty-eight.

Barley hulling mills

The horizontal corn mill could be adapted for processing other materials such as mustard and pepper, or for hulling barley (Figure 82). Barley hulling mills were introduced to the Netherlands in 1639 and by 1731 there were about 65 in the Noorderkwartier.[26] A coarse gritstone runner (Figure 83) was surrounded by a casing into which the grains were tipped (Figure 84). Inside the casing was a ring of tin sheeting in which holes had been punched so that the sharp points faced inwards making a cylindrical grater. Deep furrows in the rotating runner helped to fling the grains against the grater which removed their husks as they were 'runned' along it. The pieces of husk passed through and were later removed, while the grains were retained inside and polished. After a suitable period, the miller opened a chute and the grains fell into a hopper from where they would be graded on jog screens. For pot barley, the grains might be processed six times. Hulling stones had to revolve faster than ordinary millstones, so these mills required a strong wind.[27] Only one reference has been found to a British hulling mill. Before 1782, Sir John Dalrymple had built a mill at Cranston for grinding wheat, working pumps in a coal mine as well as preparing pot barley.[28] During the nineteenth century, some Dutch mills were converted for hulling rice but they were soon replaced by much larger ones driven by steam. Five men worked in a windmill and could process 800 tons of rice a year. In 1852, the biggest steam rice mill in Amsterdam had an output of 7,000 tons. In 1877, a new windmill would have cost Fl 22,000 and a second-hand one Fl 11,000. Wages for the workmen were Fl 2,500. A new steam mill cost Fl 110,000 with wages for men at Fl 5,000 and costs for coal Fl 2,500 but it could process 8,000 tons. In spite of the expense of coal, the steam mill was cheaper through increase in scale and the need for fewer operatives.[29]

Papermaking

One of the earliest industrial applications of windpower was to papermaking and, once again, it first appeared in the Netherlands. The expanding Dutch maritime empire of the sixteenth and seventeenth centuries needed paper for its commerce and the demand must have stimulated someone to try and adapt the windmill. It was introduced to Alkmaar and Dordrecht in 1586 but in neither of these towns have there ever been watermills because the land is too flat. While these mills could have been driven by horses, the more likely source must have been the wind. The first papermill in the Zaan area, The Goose, dates from 1605 and, by about 1740, there were probably around forty papermaking windmills there with a couple elsewhere.[1]

To make paper, old linen rags or ships sails and ropes were taken, cut into short lengths and beaten to open up the cellulose fibres and expose the fibrils.

Figure 82. The barley hulling mill, Het Prinsenhof, West Zaan, showing the typical appearance of a Dutch industrial mill (March 1992).

Figure 83. The grooves cut into a barley hulling runner stone are quite different from those on an ordinary millstone (see Figure 14) (March 1992).

This pulp was dispersed in water and lifted out on a strainer or mould as a sheet of paper which was couched off onto a felt blanket and covered by another felt. The water was squeezed out from the pile or post in a screw press; the sheets of paper were separated from the felts and hung up on ropes to dry. Later processes might consist of dipping each sheet in gelatine size to give a surface impenetrable to ink and then polishing for a smooth finish.

Water was essential because the rags had to be beaten wet; then the fibres cohere through hydrogen bonding as the water is drained off. When paper began to be made in China in the second century AD, beating was done by hammers but it is possible that the kollergang may have been used in Spain when papermaking was introduced there in the eleventh century.[2] Waterpowered stampers are thought to have been operating at Valencia in 1151 and these are the machines which pulped the rags as papermaking spread northwards from Italy into Switzerland, Germany and France. Rags would have been cut up by hand, and then taken to a series of troughs for beating by a method which originated in Italy around 1300. The stamping troughs were placed together in a row so that the shaft of a single waterwheel could be extended into a camshaft and drive tilt hammers or stampers in all of them. In the first trough, the feet of the hammers were shod with nails with cutting edges on their heads. In the next trough, the nails had ribbed heads to cut and pound the rags, while in the final one the heads of the hammers were of plain wood so the fibres were just pounded to fibrillate them. The time taken to prepare the pulp in the improved stamper troughs might be between 12 and 18 hours for the first stage and about 12 hours for the second; in the third, it was much shorter.[3] In 1541, the glazing hammer was introduced but, other than pulp preparation, papermaking remained an entirely hand operation.

The method described above must have been applied in early windmills.[4] The windshaft must have turned a vertical mainshaft which gave motion to a horizontal camshaft placed at the bottom of the mill where it could operate a line of stampers which would have occupied quite a large space. The British paper industry lagged behind that on the Continent up to the middle of the eighteenth century but in 1692 Thomas Hutton patented a scheme for driving eighty or more stampers at once by wind or waterpower.[5] Even if this referred to the

Figure 84. Barley is tipped down the hopper on the right onto the revolving runner stone. It is flung against the grater round the side of the casing through which the husks will pass into the pit on the left. When sufficiently polished, the barley will be let out through a trap door into a grader (March 1992).

Figure 85. Hollander beaters in the Schoolmaster papermill. The one on the left has the cover open so the roll with its bars can be seen, that on the right is full of pulp (June 1972).

individual hammers and allowing four or five to each trough, such a mill would have been far larger than others in England at that time. There was also an invention in 1696 by Evan Jones, Chester, of a new engine 'for supplying the Defects of Wind and Water' which could be converted to drive the hammers in papermaking.[6] It seems doubtful whether any windmills like these were built in Britain, where stampers remained the only way of producing pulp until well into the eighteenth century.

It was soon found that the wind, even in the west of the Netherlands, proved to be too irregular to provide the power needed for making the pulp with stampers and some quicker method had to be found. Some unknown Dutchman provided the answer with the hollander beater which appeared during the middle of the seventeenth century (Figure 85). The date normally given is 1670; it certainly had been invented before 1673, when a request made to the Dutch government for a patent was refused. After this it spread quickly in the region north of Amsterdam because, when the iron bars were replaced by bronze, it enabled industry there to make white paper.[7] It consisted of an oblong wooden tub with a dividing wall down part of the centre so that the rags or pulp could circulate. Metal bars were fixed parallel to the driving axle into the circumference of a solid wood drum or roll which beat the rags against another set of bars fixed in a bedplate at the bottom of the tub. To fix these bars with sufficient accuracy to beat the pulp must have taxed the engineering standards of those days and even today the bedding of the bars can take a very long time. At the same time as beating the rags to a pulp, the roll lifted them over a breast to mix them properly as they were churned round and round. The roll had to be covered with a hood to prevent splashing as the speed was sufficient to fling some back over the top where there were screens and filters. Clean water was poured into the beater to wash the rags and was removed through these screens in the hood together with the dirt.

Hollander beaters cut and lacerated the fibres, producing a different quality of pulp from stampers which rubbed and frayed the material leaving long fibres. Speed was the great advantage of these beaters and that is what the papermakers in the Zaan region of Holland with their windmills needed, in order to be able to compete successfully with the abundant waterpower of the French and Germans. In 1725, Kerfstein, a papermaker in Saxony, wrote, 'The Hollander in Freiberg furnishes in one day as much as eight stamper-holes do in eight days'.[8] By the end of the century, it was stated that it took forty of the old stampers twenty-four hours to reduce one hundredweight of rags to pulp. In the same time, the hollander beater could prepare twelve times this amount and a batch might be finished in only four or five hours.[9]

The papermakers of the Zaan area were quick to take advantage of the hollander beater and soon had devised a standard layout for their windmills which can still be seen at the sole surviving wind papermill, 'De Schoolmeester', The Schoolmaster, built in 1692 (Figure 86). The whole range of buildings is constructed of wood. Above the red tiled roofs stands the windmill itself which is a reed covered smock mill with four common sails and no fantail. The staging, where the winch at the end of the tail poles is situated, is at roof level so the sails can catch the best of the wind. The smock mill is hexagonal instead of the more usual octagon and to this shape is ascribed the ability of this mill to work in light breezes which has given it the nickname of 'De Gauwdief' (The Sneak-thief).

To what extent the interior of the mill has been modified is difficult to judge, but the basic layout must remain. The buildings are situated by the side of a canal on which barges once brought the rags and other raw materials and took away the finished paper. The canal provides water for the papermaking processes, which is pumped up to a wooden storage tank by windpower. Between the unloading platform and the body of the windmill is the rag storage house, with compartments for different types of rags. An iron beam balance can weigh out the correct proportions for the type of paper to be made. The rags are taken to

Figure 86. The Schoolmaster papermill, built in 1692, about to celebrate its three hundredth anniversary. The rags were delivered by canal, passed through sorting and cutting stages to the beaters in the base of the mill. The taller chimney heated the papermaking vat and the furthest long shed contained drying lofts (March 1992).

the centre of the mill where the main vertical shaft terminates with its bottom bearing at almost ground floor level. High up on this shaft is the first power take-off, a short horizontal shaft with cams to work a '*kapperij*' (Figure 87) where the rags are cut up into small pieces ready for the beaters.[10] While van Natrus in 1734 shows a kapperij in his drawing of an oil mill, he does not include one in his papermill but Harte, over one hundred years later, does.[11] At a lower level on the main shaft there is a second small gearwheel with two drives. One goes to a side extension of the mill where the rags may be pulped by a kollergang, possibly a later addition. It is similar to those described earlier but the trough is deeper and the sides cause the rags to fall back in again under the stones. The width of the stones gives the fibres a sort of twisting, shearing movement, helping to fibrillate them.[12] The second drive goes first to one hollander beater at a high level which is used only for mixing pulp and then glazing rolls.

At the bottom of the main vertical shaft is the large main gearwheel which drives three hollander beaters. Each beater has its own small gearwheel to turn the roll at a speed considerably greater than that of the windshaft. Layouts here

Figure 87. The kapperij for cutting rags was similar to that used for breaking up logwood for dyeing. The four vertical stampers cut the rags while a beam on the right oscillates to work the ratchet for rotating the tub (June 1972).

vary, for in some Dutch windmill books, papermills have four beaters grouped at the base of the main shaft and no kollergang. In 1849, Harte has one of the subsidiary drive shafts operating two more beaters.[13] The old mill books show that two forms of beater quickly evolved, a breaker with coarse heavy bars followed by a finisher with many more finer bars. The rags would be placed in the breaker beater first, be transferred as half-stuff to the finisher beaters and flow out of them as stuff or pulp into storage chests.

Although the storage chest at The Schoolmaster has a pulp lifter, this may be an addition installed in 1877 when a papermaking machine was added, for such a device is not shown in old drawings. Originally, the pulp would have been tipped by a bucket into the vat from where the vatman would scoop it out on his mould and turn it into sheets of paper. The coucher would lay off the sheets of paper onto a felt and the pile of felts and paper taken to the nearby screw press. The screw press still exists at the Schoolmaster and is used for the sheets of paper coming off the machine. The present paper machine has always been driven by a form of power other than wind, which varied in speed too much. The sheets of paper, after pressing, are still dried by hanging over ropes in the low drying lofts at the far end of the mill. This is a single storey building with side louvres to let the wind blow in to circulate the air and dry the paper. The height is designed to minimise obstruction of the wind reaching the sails. After drying, the paper had to be flattened, which is done at the Schoolmaster by a calender driven by a shaft made from a single piece of wood over 12.19 m (40 ft.) long. Glazing hammers were replaced around 1720 by glazing rolls or the calender, and idea of yet another unknown Dutchman.[14] The rolls on one calender at the Schoolmaster are made from lignum vitae and on the other from iron.[15] The pressure on the rolls is augmented by a weighted lever 3.66 m (12 ft.) long.[16]

The difficulty of driving a papermachine is well illustrated by the early history of the first one installed in a Dutch mill. Near to the Schoolmaster at The Fortune Mill, the four Van Gelder brothers established a new company, Van Gelder Schouten & Co., in 1838 and decided to install two machines, 150 cm (60 in.) wide, complete with steam-heated drying cylinders. They intended to drive the whole mill, including the preparation machinery, by steam engines. They ran into many technical problems with most of the equipment including the paper machine, as it was the first made by its suppliers. A second-hand steam engine instead of a new one was sent to provide the power and the steam was raised by coal in second-hand boilers too! The Fortune Mill was only a short distance away from the Beehive Mill, which produced fine white paper of superior quality. Unfortunately smoke from the chimney of the Fortune's boilerhouse adversely affected the paper in the Beehive's drying lofts so the authorities ordered Van Gelder to stop using coal and switch to the national product, peat. This proved to be impractical because it did not have as high a calorific value as coal. The windmill was employed again to drive three half-stuff beaters but the mill did not make a profit so the 20 h.p. steam engine driving the full-stuff beaters was also taken out in 1844 and those beaters reverted to windpower. A smaller 4 h.p. engine had to be retained to drive the papermachine because windpower was too irregular.[17]

It is tempting to see in early English patents schemes to introduce the hollander beater to this country. In 1692, Nathaniell Bladen patented an engine for making all sorts of paper and pasteboard from hemp, flax, linen, cotton, cordage, silk, wool and all sorts of materials much more quickly than formerly.[18] While he did not specify the source of power for driving his engines, a couple of years later, Christopher Jackson, who had a similar idea for a machine

> which dissolveth, whiteneth, and grindeth raggs, and prepareth all other materials whereof paper and pastboard hath been or may be made, in farr less tyme than the mills hitherto in use doe.[19]

envisaged using the power from either wind or water. In 1707, Peter Vallete, a London merchant, was granted a patent for his invention for making paper by a windmill and 'brasse Engines'.[20] If he did try to build a windmill, it could not have seen much use, for he was declared bankrupt in the same year.

Evidence available so far indicates that hollander beaters were not introduced to Britain until after the first quarter of the eighteenth century and windmills for papermaking after the middle. References have been found to four papermills in England where windmills were installed. These are Deptford and Cheriton in Kent, one at Northampton and one at Hull. Most is known about that at Deptford. In 1751, Josias Johannot insured with the Sun Fire Insurance Company a windmill and vats situated near the East India Dock in Deptford. Johannot was the papermaker and the proprietors were Robert Grosvenor, elder and younger, and Edward Webber. By 1753 Johannot was bankrupt but had been succeeded by Richard Mathieson. Although the mill was consumed by fire in 1755, it was soon rebuilt. While there are references in insurance documents to a mill here in November 1777, nothing later has been discovered.[21] Reference to the Northampton mill apeared in the *Northampton Mercury* in 1783 and suggests that it may have been worked jointly by wind and water.[22] The one at Cherition is dubious, for Hasted wrote in 1790.

> A little further *southward* in the bottom, between *the quarry-hills* is *Horn Street*, where the brook called the *Seabrook*. . . turns a *paper* and *corn-mill*, belonging to *Mr* Pearce, which is curious, being worked at times both by wind and water.[23]

It is not clear whether the corn mill or the papermill was driven by wind, while Hasted's maps do not show any windmill near the site. Reference to one in Hull merely states that a papermaking factory at Stepney was driven by a windmill.[24] In 1839 in Chicago, USA, Devitt attempted to operate a beater by windpower but failing here, he went to St Charles where he tried a windmill in conjunction with a waterwheel on the Fox River. This also failed and nothing more was heard of him.[25]

The handful of English windpowered papermills must be compared with the numerous ones driven by water. Forty-one papermills existed in England between 1601 and 1650. In 1700, there were around 116 mills and in 1712, the Excise authorities gave the total as 209. By 1775 this had increased to 345 and in 1800 to 417.[26] In 1785, Thomas Houghton had seen the steam engine which Boulton & Watt had installed in the oil mill of Cotes and Jarrett at Hull. Water supplies for driving his own papermill at Barrow upon Humber were insufficient and, instead of turning to windpower in a region where windmills for other industries would continue to be built for many years, he eventually ordered a steam engine for a new papermill at Sutton in Holderness close to Hull. The engine must have been running by July 1787 but there were problems with it and Houghton was declared bankrupt in 1788. A similar fate befell the 80 h.p. Boulton & Watt engine ordered by Matthias Koops for his extensive papermill at Mill Bank in London. This venture came to an abrupt end when its assets were seized in November 1802 in an attempt to recover Koops's debts from an earlier bankruptcy.[27] It was not until 1805 that the firm of Boulton & Watt installed its first successful steam engine in William Balston's new Springfield Mill near Maidstone which was to drive the beaters for the famous Whatman paper for ninety years.[28] With the introduction of papermachines starting at this time, the industry continued to expand, but it turned more and more to steam engines because water resources were limited and windmills were completely unsuitable for driving papermachines.

Mining

One of the areas in which great advances were made, to form the basis of the Industrial Revolution, was mining. Over the centuries, miners have looked to windmills to give them assistance in the varied chores of extracting minerals from below the ground. So far, no evidence has come to light for windmills actually being used in mine haulage either to raise or lower loads up and down the shafts or to pull them along galleries, although this was suggested in patents. Such machines would have been far too dangerous in practice because, if a sudden gust of wind came when the bucket was nearing the end of its wind at the top of the shaft, the mill might not be stopped in time. Equally, the wind might not have been blowing when it was necessary to commence haulage.

The first mention of a windmill connected with mining is by Georgius Agricola in 1556. He described various devices for ventilation (Figure 88). One was not really a windmill for it was merely a vertical tube descending the mine down which the wind could be diverted to take fresh air to the bottom. At the top of this tube was a barrel which could rotate. The wind passed through a hole in the side of the barrel and, to keep this hole in the eye of the wind, there was a wing or vane sticking out of the back.[1] It is included here because it shows how ancient is the use of a fixed vane, rather like a weathercock, to keep something pointing into the wind. He also described a ventilating machine with a centrifugal fan enclosed either in a square box or a better type in a drum-shaped casing. The fan could be turned by a man operating a crank and, to help, there can be seen an early form of flywheel, weights on the ends of arms. Instead, Agricola said this could be driven by a windmill. His illustration showed a post mill placed over the mine shaft. The sails were very primitive, merely half a

A—Box-shaped casing placed on the ground. B—Its blow-hole. C—Its axle with fans. D—Crank of the axle. E—Rods of same. F—Casing set on timbers. G—Sails which the axle has outside the casing.

Figure 88. Agricola's drawing of ventilating fans and a windmill driving one (1556).

dozen slats of wood nailed at one end to the main stocks. The windshaft was also the shaft for the fan which appears to have filled the whole of the buck. The air must have been forced down the supporting post of the mill which was presumably hollow. How the mill was turned to face the wind is not made clear. However, Agricola's final comments are instructive about the fickle nature of wind power.

> Although this machine has no need of men whom it is necessary to pay to work the crank, still when the sky is devoid of wind, as it often is, the machine does not turn, and it is therefore less suitable than the others for ventilating a shaft.[2]

In 1750, Parson Hale built a windmill on top of Newgate Prison, London, to drive bellows for ventilating the prison and mitigating the effects of gaol fever. This is the only other instance found of a windmill used for ventilation.

The scene moves to Britain at the beginning of the seventeenth century. Mines have to be situated where the veins or seams of minerals occur. As the miner goes further into the hill or down into the ground, he will be faced with the problem of water seeping into his workings. The easiest way is to drain this out by gravity, which is simple in the case of a drift mine running into the side of a hill, for the water can run out of the entrance tunnel. The next best plan is to dig a sough or adit below the level of the workings, to drain the water into it and so take the water away outside. However, the expense of digging an adit must have been balanced against the value of the minerals it was expected would be extracted and the capital costs of pumping equipment. Mineral seams generally penetrate deeper into the ground than the mine entrance or sough exit so the water will have to be pumped out.

There were pumps with rotary motion, like the winch with rope and bucket or the continuous chain and rag or bucket pump, as well as reciprocating pumps such as the lift or force pump. Rotary power could operate the first types quite simply through suitable shafting and gearing to turn the horizontal shaft over which the rope or chain of buckets passed. In 1659, D'Acres described the bucket gin as

> the most ancient *instrument* for water drawing (in an Artificial way) that I know of in the *world*; and there be infinite sorts of them . . . and are convenient for great *depths*, and small *waters*, and crooked and narrow pits . . .[3]

In 1588, Ramelli showed a picture of a tower mill with a rag and chain pump inside (Figure 89). The whole of the pumping apparatus and the windshaft was mounted on a central framework which turned on a pivot in the bottom of the well. The bucket chain was suspended over the windshaft and the balls of rags drew the water up from the well and emptied it into a cistern running round the top of the tower. From here the water had sufficient head to work fountains

Figure 89. Ramelli's drawing of a windmill for raising water with rag and chain pumps (1588).

in the surrounding gardens.[4] Although this is not in a mine, it does show that people were thinking of how the wind could drive such pumps. Robert Galloway pointed out the disadvantages of the rag or bucket and chain pumps many years later.

> The wear and tear was excessive; between vibration of the chains and leakage, half the contents of the water was spilled before they arrived at the top; water was constantly pouring down the pit like a deluge; and when a bolt broke the whole of the chains and buckets fell to the bottom with a most tremendous crash, and every bucket was splintered into a thousand pieces.[5]

Nevertheless, it is claimed that bucket or rag and chain pumps were used in Cornish mines until the advent of the steam engine,[6] and in other places until well into the eighteenth century.

William Pryce treated these pumps very fully in his book about Cornish mining published in 1778, and described the working conditions, which were grim. Because they had a comparatively limited height to which they could raise water, sets of them had to be installed in the Cornish tin mines, one above the other, so that men had to operate them far underground.

> The men work at it naked excepting their loose trowsers, and suffer much in their health and strength for the violence of the labour, which is so great that I have been witness to the loss of many lives by it.
>
> A rag and chain pump of four inches diameter, requires five or six fresh men, every six hours, to draw water twenty feet deep; and to keep it constantly going, twenty or twenty-four men must be employed monthly at forty or fifty shillings each man. The monthly charge of one of these engines cannot be less than fifty or sixty pounds; and they are now pretty generally laid aside on account of their great expence, and the destruction of the men.[7]

Power from wind must have seemed an attractive proposition to replace men working these pumps. But a windmill could only drive a set of these pumps at the top of a mine, where the mill itself was situated. If there were two or three sets one above the other, the windmill could have worked only the top one unless complex gearing and shafting were introduced. If the water was drained out through an adit which ran into the side of a hill, a windmill on the top of that hill could not have worked rag and chain pumps at adit level.

A windmill could have driven reciprocating pumps which might be worked by a balance lever or from a crankshaft through connecting rods. Vittorio Zonca included pictures of both lift and plunger pumps in 1607.[8] He showed men operating them through a treadwheel and crank as well as levers. The limitations of muscle power are obvious as D'Acres realised in 1659.

> The *Irrational brutes* are more strong and innocent [than men], yet subject to *tiring*: and (which oft times hurts all sides) they have not sense enough,

> to manifest (as men can) an unusual, and unwonted heaviness in the *work*, nor obstinatenesse enough to *cease* (as the *Elements* will) when it is above their strength: yet (if they be well accommodated,) they will, for one of them, perform more continuing service than four men.[9]

A man he described as willing to 'sell you his *reason* at unreasonable *rates*. . . [he] knows thereby both when and how to necessitate his service, to mitigate his pains, and aggravate his wages'.[10] Little has changed over the years! The need for an alternative source of power was obvious.

The limitations of waterpower for driving pumps are less obvious. The first problem with mining is the location, which may be well away from a suitable river. By the middle of the seventeenth century, this had been overcome to a certain extent by the use of rope or rod transmission, as again described by D'Acres,[11] but there was a great loss of power in its system. The flow down a river might be improved by storage dams and by cutting appropriate leats to lead the water into the wheel at a higher level to obtain more power but the basic problem still remained, that of having a suitable supply of water and then an adequate supply. It is this latter problem which is particularly hard to assess through incidence of drought and frost. Pryce again is instructive about conditions in Cornwall. He considered that the waterwheel was a more effectual engine than the rag and chain pump and mentioned one at Cooks Kitchen Mine 14.63 m (48 ft.) in diameter. However, although Cornwall had a high precipitation, the streams were small and scanty and many mines were in hilly regions where

> The short current of our springs from their source to the sea, prevent such an accumulation of water, as might be applied to the purpose of draining the Mines; and of course the value of the water is more enhanced. There are very few streams, which are sufficient to answer the purpose in summer, as well as in winter, so that many engines cannot be worked from May to October; which is a great loss at that season of the year, when men can work longer.[12]

Frost might diminish the supply of water and cause icing of the wheel itself so that it could not be operated. Waterpower might be almost as fickle as windpower and the miner would have to obtain the rights to the waterpower, which he would not need to do in the case of wind. There was one advantage over a windmill in that a waterwheel could be placed down a mineshaft and the water driving it could flow out along the same adit as that from the pumps. On the other hand, a windmill working lift pumps might be built in any situation where it could catch the wind without the expense of dams, leats and other watercourses.

References to windmills connected with draining mines have been discovered right across Britain, from Cornwall in the south, through Somerset to Wales, Newcastle upon Tyne and north to the Fife region in Scotland. The use of power from wind in mining may have been introduced to England by Huntingdon Beaumont. In partnership with Sir Percival Willoughby, he attempted to obtain a monopoly of coal supplies in the Trent Valley in the first two decades of the seventeenth century, bankrupting himself in the process.[13] He is credited with building the first railway in Britain, in 1604, which ran for a distance of two miles from the Strelley pits to Wollaton lane end, near Nottingham, in order to reduce the cost of transporting coal to that city.[14] Beaumont also had coal-mining interests in Northumberland and it seems clear that he had the honour of introducing the waggonway to the Newcastle area between 1605 and 1608. He took with him to the Northeast the art of locating coal seams by drilling with boring rods, which were available for sale by 1618.[15] He was willing to try out new inventions, for he is also claimed to have introduced to the Northeast improved pumping engines driven by horse, wind or waterpower. Although they were capable of raising water from below the level of free drainage, their capacity was limited to a lift of about 27.43 to 36.58 m (30 to 40 yards).[16]

Interest in, or speculation over, water-raising devices during the seventeenth century can be discovered in patent specifications. Seventeenth century entrepreneurs looked to the Continent and particularly the Dutch for bright ideas which they could exploit through being granted the sole rights both to protect the product and to manufacture it. The government wished to encourage new industries and a royal warrant in the form of a patent would help to smooth the projector's path.[17] It was against such speculations that Walter Blith inveighed in 1649.

> We had some Mountebancks abroad that have held out specious pretences of wonders, as many Engineers have done in drawing water, or drayning Lead-mines, Tin, or Cole-Mines, and to that purpose have projected Engines with double, treble, and fourfold Motions; conceiving and affirming, every Work, or Motion, would multiply the ease in raising the water, but not considering that certainly it must multiply the weight and burthen thereof, and also put such an Impossibilitie unto Tackles, Geares, and Wheels for holding, that all would flie in sunder, because of the great strength is required to move the same, mistake me not, I do not here reprove the use of Engine Work, a good Engineer is a gallant and most usefull Instrument in a Common-wealth, and they have principles most able to make the best Husbands and Improvers, I onely warn you of Imposters: Engines are most necessary, and easeth all our burthens.[18]

In spite of Blith's warnings, various hopeful inventors took out patents which mention windmills and improvements to mine drainage. Even before 1600, Daniel Houghesetter was interested in draining mines with his engine which would raise 'waters from anye place whatsoever from low to high'.[19] In 1674, John Johnson had a new manner or motion for windmills 'by which may be performed the raising of water to great heights, the draining of mines and drowned grounds, and other things'.[20] No further details are given either about the windmill or the pumps. Nathan Heckford, ten years later, patented a type of horizontal windmill 'for draineing of lands, pitts, mines, and many other uses, att cheaper rates than ever', but again no details are given.[21] Finally, in 1692, although Thomas Hutton primarliy envisaged making paper with his windmill, he included 'another new-invented engine apperteyning to it, for the raiseing great quantities of water, singularly usefull in the draining coale mines, lead mines, and other like dreynings or water riseing'.[22] While it is impossible to say whether any of these projectors erected a windmill, other people did so.

In 1698, Sir Humphrey Mackworth tried to improve his colliery at Neath, Glamorgan, by constructing a railway across the main road between that town and Cardiff. The grand jury at Cardiff declared that this was a 'nuisance' and part of it over the highway was torn up. He employed his coal to smelt copper and, to convey the coal cheaply to the harbour, he built some wagons with sails, whereby 'when any wind is stirring one man and a small sail does the work of twenty'.[23] He also set up windmills at his mines and the poet Yalden wrote about him.

> The winds, thy slaves, their useful succour join,
> Convey thy ore, and labour at thy mine.[24]

An agreement has survived for the construction of a windmill in the North Somerset coalfield at Wraxall. It was dated 30 December 1700 and John Bryant of Portbury was to build

> an Engine to go with ye winde in manner as a windmill in ye Parks called the South Parke in Wraxall to draw water out of the Pitt now begunne there with the help of Pumps from the depth of Twelve fathom that shall carry off a Shoote of water to ffil a Bore of Six Inches Diameter which sd Engine shall shall bee reddy to worke and carry off the whole shoot of water belonging to the coaleworks as the sd coaleworks shall from time to time bee wrought by the 28th day of February next.[25]

It is interesting to note that this was not the first windmill of this type in that area, and also how quickly John Bryant was expected to erect it, in just a couple of months.

Celia Fiennes, that intrepid traveller, who visited Cornwall in 1695, was mistaken when she wrote, 'Saw not a windmill all over Cornwall and Devonshire, though they have wind and hills enough, it may be too bleak for them'.[26] Not only had windmills been used for grinding corn in both these counties for centuries before her visit but there were also some connected with mines. A map, drawn by George Withiel in the 1690s, covers the Hartley estate at Tremenheere and Terillion at Wendron, an area above the headwaters of local streams. It has at the northern extremity of Tremenheere, bordering on 'Rose-litton' (Roseliddon), a note that 'upon this piece of ground is a great tin-work'. Two lodes are marked with a windmill evidently for pumping on one.[27]

To what extent windpower was used at this period remains a mystery but this windmill suggests that people were looking for better means of powering their pumps. Thomas Savery patented a steam engine in 1698 which he demonstrated to the King and the Royal Society in the following year. In 1702 he published his book entitled *The Miners Friend* because he hoped to exploit this potential market. However, for draining deep mines, his engine proved to be impractical, the materials from which it was made being unable to withstand the high boiler pressures necessary, and there were other technical problems.[28] There is no record of his design ever being tried in a Cornish tin mine at this period although one may have been placed in a coal mine at Willingsworth near Dudley.[29]

Thomas Newcomen's engine, erected near Dudley Castle in 1712, was able to cope with raising water from deep mines by using lift pumps operated by a lever beam, which formed the basis for later improvements by John Smeaton and James Watt. Newcomen is known to have visited mines in Devon and Cornwall where he supplied all sorts of ironwork and tools. It is possible that he may have erected an earlier engine there, for Joseph Carne in his paper on the history of copper mining states that the first engine in Cornwall was erected at Wheal Vor tin mine in Breage where it worked from 1710 to 1714.[30] The owners of Savery's patent claimed that it covered Newcomen's engine, too, and so Newcomen had to join these proprietors to erect his engines. Over one hundred had been built before the patent expired in 1733. Steam engines based on Newcomen's design were to supplant other forms of power for draining mines and windmills were often the first to be affected.

Newcomen himself had some interest in windmills for in 1725 John, the elder son of his former partner John Calley, was involved with attempts to patent a new type in Holland. In 1724, John Brent had patented a windmill 'far exceeding all wind engines hitherto practised whose fanes move horizontally'.[31] Newcomen wrote to Lord Chief Justice King asking him to use his influence with Sir Matthew Decker to get an introduction in the Netherlands for his nephew, George, to obtain a patent. This was necessary because John Calley, who was charged with the mission, had died in that country. A Dutch patent was granted to 'Johan Brent and Johan Celly' on 24 January 1726.[32] To what extent Newcomen was involved with windpower remains a matter for speculation but it is significant that British technology was being transferred to the Continent by this date.

In 1738, John Kay was described as 'Engineer' at Bury in Lancashire when he patented a horizontal windmill.[33] The wind part had eight arms with sails hung on moving frames opening and shutting like a door so that the wind could be received from any direction without having to turn the body of the mill itself. Kay described the disadvantages of raising water from mines or deep pits with current apparatus. When two buckets were used on either end of a rope, the horse had to stand still while one bucket at the top was emptied and then had to reverse to bring up the other. With a chain of buckets, there were frequent breakages. He patented methods of overcoming these problems. To what extent Kay was involved with mining remains unknown but Edmund Lee, who patented the fantail in 1745, came from nearby Wigan, another mining area in the locality.

Moving north to Scotland, in 1598, Gavin Smith and James Aitchison (goldsmith to King James) were granted a patent for a mine-draining device to be powered by wind, water, horse or men.[34] In 1661, the Earl of Wemyss was empowered to sink a coal-mine at Cameron, Fife and to erect wind and watermills and anything else necessary for winning and transporting coal.[35] George Bruce tried windmills at Culross during the seventeenth century, without conspicuous success.[36] In 1708 a plan was drawn up for draining some collieries with windmills in the Fife area. So backward was the state of engineering in the region that there was only one person capable of putting it into effect: John Young, a millwright of Montrose who had been sent to Holland by that town to inspect machinery. If his services could not be obtained, it was suggested that advice should be sought from the 'Mechanical Priest of Lancashire'.[37] Who this was, would be interesting to know. Windmill installation continued in the first half of the eighteenth century. The Earl of Mar, proprietor of Alloa collieries in Clackmannanshire, wished to introduce better pumps and in 1710 sought the advice of George Sorocold who in 1702 had been responsible for installing the waterwheels in the first silk mill built by Thomas Cotchett at Derby.[38] He recommended lift pumps worked by levers instead of the chain and bucket type. No millwright capable of constructing such a machine could be found in Scotland and a chain and bucket pump had to be built.[39]

Lift pumps were used in later windmills in Fife. For example, John and Alexander Landale leased Balgonie coal from the fifth Earl of Leven. In 1732 they were using both drainage waterwheels and a windmill which was lifting water from a depth of 14 fathoms (25.6 m; one fathom = 6 ft.) with eight pumps.[40] But it is from the Rothes manuscripts at Kirkcaldy that a great deal can be learnt about the type of mill and the costings. These documents include a drawing which shows a tapering tower mill sectioned to reveal the machinery inside (Figure 90). The mill is built to one side of the mineshaft. The cap is conical with a tail pole to turn it into the wind. There are four common sails on a windshaft with a large brakewheel, although no form of brake is illustrated. The gears are the trundle or lantern pattern pegged into their respective wheels at only one end. A vertical mainshaft takes the power down to another pair of gearwheels to turn a short horizontal shaft within the mill. At each end of this shaft is a crank linked by connecting rods to the inner ends of pivoting beams. At the outer ends of these beams are arch-heads round which chains run to hang down the pitshaft to the pumps. A subsidiary drawing at the head shows two

Figure 90. A windmill used in Fife, Scotland, for raising water from coal mines with lever pumps (Rothes Manuscripts).

pairs of pumps in double lifts with wooden pipes conveying the water up the shaft. There survives a detailed estimate for the construction which cost £53.7.0. The amount of water which it raised is also given.

Calcul of this machine	Hogshead	Gallon	pint
Ye weight of water in 20 fathm of 7 inch pumps in 1910 p.	238.	50.	3
at on stroke with each leaver discharges allow a 4 foot stroke,		16.	0
When ye machine goes 17 times round per minute	4.	0.	0
and per houer	240.	0.	0[41]

There is another estimate for a windmill for Lord Rothes, dated 1738, which includes dimensions. The sails were 18.28 m (60 ft.) long, with the windshaft 3.66 m (12 ft.) by 76.2 cm (2 ft. 6 in.) square. The brakewheel had forty teeth and the gearwheel on the vertical shaft to mate with it had twenty. This vertical shaft was 11.58 m (38 ft.) long and again the gearwheel at the bottom of it had twenty teeth mating with one of forty on the crank axle. The cranks themselves had a radius of 60.96 cm (2 ft.). The tail pole or 'rudor' was 10.1 m (33 ft.) long 25.4 cm (10 in.) square. In this instance the pumps were 48 fathoms (87.78 m). The lever beams to the pumps were 7.62 m (25 ft.) long and 40.64 cm (16 in.) square with arch-heads 2.13 m (7 ft.) long. The annual costs of running such a mill are given too.

Acc of the annuall charge of the Wind-mill at Drougan, by Mr Landale, 1737.

Estimate of the Yearly Expense of Keeping going a wind-mill.

	£ s d
To two men to attend the one in the day and the other in the night at 13½ d & 6⅔ d	26. 0.0
To leather in the pitt being kept clear of coal	1.10.0
To 72 yrds sail-cloth at ¼ d pr. yrd.	4.16.0
To ropes	15.0
To a smith for making hoops & forelock for the boxes and other incident	1. 0.0
To grease	1.10.0
	£35.11.0
Taking of the man in the night is	8.13.4
The expense will be	£26.17.8[42]

The cost of the materials needed to build the mill at Drougan was £152.0.5. This included freighting the pump barrels from London and the brasswork from Kinghorn, both to Leith. The cranks, probably for a different windmill, were made at Newcastle. This reveals how backward was the state of engineering in the region, with most of the metal parts having to be shipped in from elsewhere. The cranks with freight cost £23.7.3½ and the pump barrels £15.18.9. Brasswork cost £8.2.6, other ironwork £15.11.4½, timber £30.0.0 and the millwright's work £50. Building the windmill tower cost £12.18.8 but it is not clear whether this covered the materials. The estimate for another windmill with 16.46 m (54 ft.) sails for drawing water up 30 fathoms (54.88 m) was only £115.4.0 for the millwright's work.[43] Yet another estimate for erecting a windmill and sinking the shaft for a pit at Strathore worked out at £402.0.0 including £50 for the pitshaft and £60 for 'deals for the sinke', presumably the tubbing round the shaft.

These figures are comparable with a 6.4 m (21 ft.) diameter waterwheel and pumps put up for the tenth Earl of Rothes at the same time by Stephen Row, who was also responsible for at least one windmill. Row was overseer of all the collieries in the district and probably came from England. The waterwheel pump was capable of performing nine strokes per minute and, with its double beams, raised 185 hogsheads (11,655 gallons) of water per hour. It was erected at Strathore near Clunie and cost over £200. Roughly fifty years later, a waterwheel, beams and pumps might cost up to £250.[44] The Newcomen engine installed at Edmonstone Colliery in Midlothian in 1725 cost £1,007.11.4 in materials alone plus £80 a year for eight years for a licence without the cost of the engine house. The engineers who erected the engine were also to maintain it and were to be paid £200 a year plus half of the clear profits of the colliery. An atmospheric engine bought for colliery drainage by Boultbee of Swannington in 1760 cost £1,500 and Scottish examples cost about the same.[45] Therefore it would seem reasonable to allow around £1,300 per atmospheric engine to cover engine, pumps and engine house.[46] Annual running expenses are unknown. The lesser capital needed for a windmill was very attractive, as was its much simpler technology. As long as the sale of coal remained localised, the capital cost of a steam engine almost certainly made it an uneconomic proposition for small pits and helps to explain the interest in windmills. It was only later, when coal markets expanded, that steam power for pumping became a practical reality.

The Rothes manuscripts give the performance of a windmill 10.97 m (36 ft.) high with 10.36 m (34 ft.) sail length and width 2.45 m (8 ft. 3 in.).

> With a moderate Galle of wind the milln will draw 24 hours off our water in ffour hours time and so in proportion when she stands longer or shorter time but that depends much on the constancy of the wind and care of the Keeper. This milln performs her work very well as yet but I presume she would have been better if the blades had been two or three foot longer

> and a little broader. But the man was affraid of the storms we have ffrom the West seas which are seldom so violent in the East countrie and more constant. She has stood her quarantine this winter and is all saffe as yet. She was made every way strong which probably would not have been the case iff the builder had been to provide all the materials and conffined to the estimate.[47]

The Glasgow coal owner, John Gray of Carntyne and Dalmarnock, was persuaded not to purchase a Newcomen engine, owing to the expense, for his important Westmuir Colliery near Glasgow in 1737 and chose a windmill which raised water with tolerable efficiency 'until the windy Saturday, 13th January 1740, when it was blown to pieces, and never again refitted'.[48]

Sir John Clerk, after a lifetime spent in this land of gales, could not recommend investment in windmills through 'Want of Wind, which one would not readily suspect in a Country like Scotland'.[49] Though a few hours' pumping a day would clear the water accumulated over a twenty-four hour period, the wind might not blow for a fortnight during which time the colliery would be flooded. Robert Galloway commented that windmills,

> though they were powerful their action was found to be too intermittent; the mines being drowned and all the workmen thrown idle during long periods of calm weather.[50]

Yet, as late as 1782, Sir John Dalrymple installed not only a water engine to drain his coal-mine at Cranston, Midlothian, but also a windmill which worked pumps, ground wheat and prepared pot barley.[51]

Dr William Boase in his Excursion Note Book of 1752 may point to one problem associated with draining mines by windmills.

> Saw a Wind Engine, like a Windmill with fanes, two miles before St. Austle – but when the water was drawn out they could not check it. Whilst it work'd it drew more than the Fire engines, but now idle – It wants a counterpoise when there is no water. Query, if an occasional bob to be fix'd on and then taken off would not answer the end proposed.[52]

Water would be seeping into a mine twenty-four hours a day and must have been collected in a sump before it could be pumped out. The account of the windmills in Fife observed how this water might be raised in only four hours of actual working. While there was water in the sump, the load would be nearly constant but, as soon as the sump was drained, the pumps would suck air and had to be stopped quickly. A steam engine could be worked whenever it was required and the engine-men must have learnt to judge when it would have nearly emptied the water and so stopped it before any damage could occur. A windmill had to be worked whenever there was sufficient wind, whether the sump was full or not, and it must have been much more difficult to judge how much it had pumped out. Also, sudden gusts would make it pump more quickly, with the danger of it sucking in air when the sump level was low. A low level in the sump and a sudden gust might cause the windmill to run away catastrophically. This happened in 1797 to Richard Trevithick with the windmill he installed at the Ding Dong Mine in the parish of Madron, north of Penzance. Sometimes it went so fast that the men could not stop it. Some sailors came from Penzance and made a plan for reefing the sails, presumably without much success.[53] When Trevithick was being consulted about replacing windmills with steam engines for draining parts of Holland in 1828, he designed a modernised form of rag and chain pump to be worked by a steam engine.[54]

How to improve windmills for mine drainage occupied inventors throughout the eighteenth century. Anthony Parsons in 1734 linked a form of horizontal windmill to a waterwheel for communicating

> motion to three large beams of wood, like unto those of the fire engine which are moved from the centre upwards; from each end of these beams goes rods of deal to the bottom of the pit or mine and forces the water up through two pipes of lead to the height of one hundred and fifty or two hundred yards.[55]

If he thought that he was patenting a new development with these pumps connected to windmills, he was a little late, but the attempt to transfer technology from steam to windpower is worth noting.

Some inventors came from Devon and Cornwall. Richard Langworthy, a 'Chirugeon' of South Brent in the County of Devon, intended to use his horizontal windmill to 'sett stamping-mills, windmills and other engines for drawing and extracting water, metal ores, etc. to work'.[56] It is doubtful whether any windmill was ever used for winding. In 1785, Christopher Gullett of Exeter also hoped to use his 'new invented eolian engine for working pumps, ginns, whyms, cranes, or other machines by the force and action of the wind and the power of compressed air'.[57] He gave the reason for his invention which

> was the bad behaviour of a person who had contracted for drawing the water from a particular part of the said mines (lead mines in the parish of Beerferris) by a horse engine and water barrels, and from a dread of listening to the recommendation of a fire or steam engine, on account of its great expence. The first idea of it was from looking at a shuttlecock.[58]

The cost of steam power remained prohibitive for small enterprises to the end of the eighteenth century and in places well into the next.

It is not clear from the specification whether John Rowe of Perransands drove his mill for grinding ores by wind or water.[59] But in 1787 Benjamin Heame, a merchant from Penryn in Cornwall, took out a much more significant patent,[60] concerned with regulating the power of the sails. His windshaft was bored hollow to take a control rod which protruded at the outer end beyond the shaft so that ropes or braces could be attached to the sails. At the inner end, a weight hung down over a fusee. When the wind became too strong, the weight could rise, allowing the sails to spill the wind. As the wind diminished, the weight fell back to its normal working position, turning the sails to the wind again. This form of automatic regulation, while ingenious, was too complicated and cumbersome to be of practical use.

The prospect of economy was always the prime concern of Cornish miners which led to further attempts to use windpower in the nineteenth century. In 1824 some speculators were building a windmill at St Agnes for pumping water from their works. In 1848, a windmill was erected on the summit of Kit Hill near Callington to drain a mine nearby; a mine that was small and, being in granite, was relatively dry. In 1849, there was yet another attempt to use a horizontal windmill at Wheal Widden in St Just.[61] However, all of these were probably forms of speculation and a more correct picture may have been given by William Pryce who, in 1778, never mentioned windmills in his book on Cornish mining.

One place where windpower for mine drainage does seem to have been more successful was Parys Mountain on Anglesey, North Wales. On the top of the hill, amidst that extraordinary lunar landscape of dereliction left by the search for copper, stands the tower of a windmill. Anglesey is a good region for windpower and the Parys Mountain, being one of the highest parts of the island, must be one of the best sites there. The copper vein was tapped in 1775, just in time for the boom created by the American War of Independence. Exploitation was rapid and, by 1780, 3,000 tons of metallic ore were being produced a year. A map of 1785 of the Mona mine shows a windmill in use for pumping.[62] Wooden pumps hewn out of large oak trunks were used because the nature of the water corroded iron. In 1878, the last tower mill built anywhere in Wales was erected here (Figure 91) with five sails and assisted a steam engine.[63] It was still working in 1901.[64]

Yet, when the large number of all sorts of mines is considered, the number of windmills at them is insignificant. By 1733 there were over one hundred Newcomen engines and this number continued to increase throughout the eighteenth century at both coal and metalliferous mines. Between 1775 and 1800, the partnership of Boulton and Watt erected 164 pumping engines based on their more efficient design with the separate condenser.[65] Most of these were in the tin and copper mines of Cornwall. To what extent traditional windmills were used in mining abroad remains unknown. There was at least one in the Swedish iron ore mining regions. Boulton and Watt were consulted about replacing it in 1788 with a steam engine and a picture around 1805 shows a smock mill with the arch-heads of lever pumps outside it. In the background are a couple of steam pumping engines and what appears to be a post mill in the distance.[66] No other continental windmills associated with mine pumping have been traced.

Figure 91. The windmill tower on top of Parys Mountain, Anglesey, built in 1878 to pump water from the copper mines (August 1991).

Waterpowered sets of stampers for pounding metal ores were shown in Agricola's book in 1556.[67] In the 1690s, the Earl of Hopetoun established a windpowered crushing and smelting mill at Leith near Edinburgh.[68] Richard Langworthy in 1749 expected his patent windmill would 'be very useful to persons occupying windmills and stamping mills, and also on shipboard'.[69] No date has been given for the round tower which was once a windmill at Warmley and contained ore-crushing machinery for the Bristol brass industry.[70] The *Mining Journal* in 1870 wrote

> We are glad to see an indication, or rather promise, of fresh economies in the revival of the proposals to use wind-power in mines. Cornwall is quite as windy as Wales, all things considered . . . There are, by the way, several indications that the use of wind-power in tin-dressing was not unknown to the 'old men', both in Cornwall and in Devon, particularly in connection with the old 'crazing mills', to which it was easily applicable.[71]

Windpower would have saved scarce water resources and expensive coal but, from lack of records, its use must have been infrequent.

So far no instance has been discovered of a windmill driving machinery in an ironworks. Blast furnace bellows were operated by camshafts but, for such a task, windpower was probably too fickle and furnaces were normally situated at the bottom of hills. In 1728, an Englishman, John Payne, patented grooved rollers for producing round bars of iron which he proposed to operate by a sort of windmill with the draught in a building created by the rush of air being drawn in by large fires.[72] This was impracticable and the vagaries of the wind account for it not being applied to drive forge hammers. The smith needed to take his red hot iron to the anvil and hammer it immediately before it cooled. He would not have lacked colourful language if he had tried to work his wind powered hammer just as the wind died to a calm.

Textiles and agriculture

In those industries which formed the basis of the Industrial Revolution in Britain, farming, textiles, mineral extraction and metal working, windpower made little impact.

The textile industry

Windpower was used in Belgium and Holland to scutch flax and hemp. Linen is made from the bast fibres of flax plants. The plants were pulled up and dried after which bundles of the stems were put into pits and soaked or 'retted' in mildly acidic water to soften the woody pith surrounding the fibres in the stem and make it brittle after drying once more. The pith could be removed by bending the bundles to break it up and then beating or scutching. During the seventeenth century, attempts were made to mechanise these processes and English patents were taken out by Abraham Hill in 1664[1] and Charles Moreton and Samuell Weale in 1692.[2] A waterpowered 'flax' mill was in operation in the north of Ireland by 1717 and a 'lint' mill was let in Scotland in 1726.[3] However, in Scotland, the Board of Trustees decided to see if these machines could be improved and sent James Spalding to Holland to collect models of the most efficient Dutch types and learn their techniques. By July 1729, he had returned and fitted up a machine in Edinburgh. First trials with horse and waterpower were not very successful and it was only after an experienced millwright, Andrew Mitchell (possibly Andrew Meikle), had been called in that the machinery worked satisfactorily.[4]

To break up the pith, the bundles were passed between horizontal fluted rollers which bruised the flax. To remove the pith, wooden blades, like blunt knives, were placed around the circumference of a wheel and were rotated just the other side of an upright board with a slot in it. The bundle of flax was stuck

Figure 92. The hollow post mill for scutching flax at Kortrijk, Belgium (August 1980).

through the slot and was hit by the blades which effectively knocked out the pith. When one end of the bundle had been cleaned, the operative turned it round and finished the other. In one layout, the driving spindle was vertical and the workers stood around a circular casing with vertical slots in it.[5] The more general form had a horizontal shaft with a series of wheels along it with the blades fixed radially. The workers stood behind a vertical screen beside each wheel. In Scottish lint mills, both bruising rollers and scutchers were driven by waterpower.

A flax scutching windmill has been preserved at Kortrijk in Belgium (Figure 92). The large top of the post mill, with thatched roof, wooden windshaft and some wooden gearing, might indicate an early date for such a mill. However, an iron main shaft through the hollow post and the iron framing of the scutchers point to the eighteenth century or even later. Here the initial bruising was carried out by hand-turned rollers with, in the large round house, only the horizontal-axle scutchers driven by wind. The size of the blades on the scutchers must have made the workers careful not to step backwards into those of the stall just behind (Figure 93). Scutching would have been a task which could be fitted in among other duties on the farm whenever it was opportune or, in the case of a windmill, whenever there was a suitable wind. There would seem to be no optimum speed for this process which could not have needed much power once the mill was turning and variations in speed would have been of little consequence. In Scotland, the improved scutching mills proved to be faster and just as effective as doing the job by hand and the same comments would seem to apply to this windmill. In the Netherlands, hemp was prepared in windmills. The illustration in one eighteenth century mill book shows a row of nine vertical stampers with plain feet for pounding the hemp. This was a dusty process and photographs of later mills show large windows at the floor where the hemp was being beaten so the dust could be blown away. In 1731, there were over twenty hemp mills in the Noorderkwartier.[6]

While these processes were suitable for being powered by windmills, the same cannot be said of spinning. Among those listed as 'Industrial Windmills in Norfolk', there is one in Norwich on Carrow Hill, with the date 1800, driving cotton machinery; and another at Diss, dated 1838, driving a yarn spinning factory.[7] The fact that no other references have been found confirms the unsuitability of windmills. In spinning mills, power requirements might rise and fall

Figure 93. The stalls and blades on the scutching rotors at Kortrijk (August 1980).

as machines were put into or taken out of gear when creels became empty or bobbins full. The power from wind would not match these variations so that the speed would alter. But for spinning, a constant speed was required to avoid gearwheels hunting and causing the machines to deliver thick and thin parts in rovings or yarns.

Samuel Unwin was a hosiery manufacturer and merchant under the domestic system, who must have wished to secure his supplies of cotton yarn for he built a spinning mill at Sutton in Ashfield around 1770 based on Richard Arkwright's system. It was first powered by horses which were later replaced by a 7.32 m (24 ft.) diameter waterwheel when the mill may have been extended. The River Idle on which the mill was built is sluggish and its water supply soon proved inadequate, so Unwin built a windmill on the roof of the mill at its southern end. This was used to pump back into the upper reservoir water which had flowed down to the lower one after driving the waterwheel. The windmill also proved inadequate and by about 1790 a steam engine had been added. A promissory note with the incomplete date 181– has a picture showing a tower mill with four sails standing above the middle of the castellated Gothic façade.[8]

To remove the natural grease in wool and any that might have been added for spinning and also to make the woven fabric more compact, woollen cloth had to be fulled. There was a waterpowered fulling mill in Normandy in the eleventh century and Italy had them in the twelfth. The earliest known fulling mills in England were situated on lands belonging to the Knights of the Temple in Yorkshire and the Cotswolds, which were mentioned in a document of 1185 when they must have been at work for some time. Scantiness of twelfth century records makes it unlikely that these were the only fulling mills then at work and references multiply in following years. In the reign of Edward I, records have been found of between 120 and 130 mills before 1327.[9] It was the spread of such water-driven mills that gave rise to the notion of an industrial revolution in the thirteenth century and it is interesting to note that this period saw the spread of the corn-grinding windmill too.

A primitive form of fulling mill has been preserved at the Romanian Open Air Museum in Bucharest.[10] The axle of the waterwheel has been extended to drive a pair of stocks with two hammers in each trough. Each hammer was lifted twice in one revolution of the axle. The hammers themselves have stepped heads and were lifted directly by the cams on the axle through tappet pegs driven in near the bottom. The hammers are slightly inclined from the vertical and slide up and down in a frame so they need no horizontal shafts. This inclination means that the cloth could be given a sideways blow, which must have helped to turn it round in the trough to ensure that it was beaten evenly throughout its length. These Romanian stocks may give a clue to the origin of an early layout similar to that in Dutch windmills. The speed of the hammer blows would appear to be of little consequence, and presumably the cloth would come to little harm in the stocks if the wind were to fail. However, no windmills of this type survive today, suggesting that the design was not very satisfactory.

Illustrations in Dutch windmill books show rows of troughs placed within the body of the mill. Like the Romanian stocks, the hammers were lifted directly by a camshaft, though here the tappets were placed near the tops of tall hammers. This enabled the camshaft to be situated high up in the body of the windmill where it could be driven by a suitable right-angle drive from the main shaft. The hammers were simple baulks of timber set vertically in a framing. Their feet were stepped but, with only a vertical movement, they could not have given much rotation to the cloth and this could be a possible explanation of why such mills disappeared relatively quickly. In one mill picture, there were five troughs, each with a pair of hammers, on one side of the driving gearwheel and three troughs on the other. A subsidiary shaft has four troughs, making twelve in all, so it was a large establishment.[11] The same layout is reproduced in a book of 1849, so some fulling mills may have been still working then.[12] In the region of North Holland, these fulling mills were reasonably successful compared with Tilburg in the south where they were tried, but the wind was insufficient. No mills like these have been found outside the Netherlands. Presumably in Britain there were enough waterpower sites in areas of woollen manufacture for there to be no need to use windpower for fulling. John Smeaton demonstrated in 1759 how the efficiency of waterwheels could be improved and so increase the amount of waterpower available.[13] With the introduction of iron construction after about 1800, larger wheels could be built, but even all these improvements could not meet the insatiable demand for power generated by the Industrial Revolution.

Waterpower was preferred to steam for spinning mills until long after 1800 and so the cotton industry spread over Britain in order to find the necessary water resources for mills such as New Lanark, Catrine, Cromford, Belper, Styal and Holywell. The cotton industry was in the forefront of technical innovation through the inventions of Richard Arkwright and others. The system of roller spinning, which Arkwright patented in 1769, could be power-driven. To supply cotton to his spinning machines, Arkwright had to invent a carding engine and in his 1775 patent he specified a complete power-driven factory system for spinning cotton. His first mill at Nottingham was powered by horses and, although he moved to Cromford to a waterpowered mill, many other people built horse-powered spinning mills, which reflects the lack of suitable waterpower sites in textile areas such as Lancashire. From about 1780, the cotton industry began to expand dramatically. In 1783, Arkwright built a cotton mill on Shude Hill, Manchester, which had to be powered by a Newcomen type steam engine recycling water for a waterwheel because there was no river in the vicinity. The cotton-spinning industry was one market for the rotative steam engine developed by James Watt and its importance may be seen in the numbers supplied to textile mills. Up to 1800, the partnership of Boulton and Watt built 496 engines, 164 pumping engines, 24 blowing engines and 308 engines driving machinery. In this last group, 110 were supplied to textile mills and of these 92 were for cotton mills.[14]

Agriculture

Although small windmills were employed for land drainage, they were seldom used on farms for such duties as butter churning, chaff cutting or threshing. In Scotland, Andrew Meikle patented his threshing machine in 1788,[15] which could be powered by horses, water or wind. While buildings to house horse-wheels and even a few steam engine chimneys can be seen on Scottish farms and elsewhere in Britain, the use of wind was infrequent, possibly because horses were always available and were more dependable. According to figures for Midlothian, water and windpowered threshing mills were cheaper to operate in 1811 than those driven by steam. Waterpowered mills cost £150–£160 each to install compared with windpowered ones at £450–£470.[16] High cost may partly account for the scarcity of windmills driving threshing mills in East Lothian in the early years of the nineteenth century and their total eclipse by 1853.

Power	Number of Machines *c.* 1839	Number in 1853
Steam	*c.*80	158
Wind	7	–
Water	30	81
Horse	269	107
Total	386	346[17]

The empty tower of a windmill beside a barn at Hilton Farm, Ellon near Aberdeen, still survives. On the Isle of Man, there were some farms which had windmills to work barn machinery,[18] and this was one use to which the annular vaned American style windmill was put around 1900, but never very frequently.

Overseas in the eighteenth century, tower mills were built in Jamaica to power vertical crushing rollers for sugar cane. On other small West Indian islands with limited water resources, such mills were quite common.[19]

Watt did not perfect his rotative steam engine with the parallel motion until 1784 and then held the vital patents with that one and his separate condenser until 1800. These early engines required heavy capital expenditure, were large and cumbersome and were costly to run, so did not compete at first with windmills, particularly for agriculture. However, after 1800 this began to change when the field was opened up. Rival types appeared and cast iron construction with easy reproduction of standard parts must have reduced costs. Richard

Trevithick had been experimenting with high pressure steam from about 1800 and his engines were more compact and lighter than the low pressure Watt condensing type. Trevithick designed small engines suitable for use on farms. In 1812, he supplied one to Sir Christopher Hawkins, Bart., MP, Trewithen, to replace the cattle mill for threshing corn. Trevithick found that 2 bushels of coal costing 2s. 6d., did the work of four horses costing 20s. Another of his engines worked until 1879.[20] Small steam engines were introduced on farms and estate workshops, particularly in areas with cheap coal supplies. Erddig House near Wrexham, built on top of coal seams, has a little vertical steam engine of about 1850 for driving the sawmill and other machinery in the estate workshop. By this time, the portable agricultural steam engine was being manufactured with a boiler, cylinder and driving gear mounted on wheels. Nathan Gough of Salford produced one in 1830 but the type did not really assume a practical form until shortly before 1850.[21] All these steam engines began to threaten the position of windmills not only in agriculture and industry but even in corn grinding, for portable engines could easily be coupled to mill gearing and drive millstones in periods of calm.

CHAPTER 8

The demise of the traditional windmill

Traditional windmills can still be seen scattered across many parts of northern Europe but very few have any profitable commercial function. The windmill has been a long time in dying, which shows how well suited it was for its role. Many met their end through fire. Standing in prominent positions to catch the wind with their sails reaching up into the sky made them prime attractions for lightning strikes in thunderstorms. In a sudden severe storm, friction in the brake might generate so much heat trying to stop or hold the sails that sparks might fly off the wooden parts. Being in remote places, it would take time to summon help and in any case supplies of water would be scarce on tops of hills so, once the wood was well alight, the inferno was difficult to extinguish. Unless well protected, wood will soon start to rot even in a working mill and, once begun, deterioration progresses at an ever increasing speed. As windmilling became more and more marginal, so there was less money to keep the mills in repair and, once something broke, restoration became uneconomic. Often it was the stocks which rotted but the mill might be worked for a time by some alternative power source. As small-scale milling became unprofitable, all production ceased, eventually leaving the mill an empty shell.

In England, we have seen that the industrial windmill was never a really viable unit and reference to them dies out in the nineteenth century. The number of corn mills seems to have peaked around the 1830s but even at that time people foresaw the end of traditional mills. James Simmons, a papermaker at Haslemere, Surrey, wrote in his diary in January 1841, 'George Newman is wishing to take a windmill near Chichester. I wished to dissuade him from it.' He was quite right, for his diary entry of 12 March 1844 stated,

> Mr. Wyatt brought word that the Compton windmill was burnt down yesterday morning. George Newman rents it, he was there at the time, the winds very high, he tried to stop it, but could not and the friction was so great that it ignited and burnt it down. He was insured and so was the Landlord.[1]

William Fairbairn's attitude was probably typical of most engineers by the middle of the nineteenth century, when he wrote in 1861:

> It is within my own recollection that the whole of the eastern coasts of England and Scotland were studded with windmils; and that for a considerable distance into the interior of the country. Half a century ago, nearly the whole of the grinding, stamping, sawing and draining was done by wind in the flat countries, and no one could enter any of the towns in Northumberland, Lincolnshire, Yorkshire, or Norfolk, but must have remarked the numerous windmills spreading their sails to catch the breeze. Such was the state of our mechanism sixty years since, and nearly the whole of our machinery depended on wind, or on water where the necessary fall could be secured. Now both sources of power are also abandoned in this country, having been replaced by the all-pervading power of steam. This being the case, we can only give a short notice of wind as a motive power, considered as a thing of the past.[2]

Yet this statement by Fairbairn has to be accepted with caution because new windmills were still being built on the eastern side of England in places like the Humber estuary and even round to Great Yarmouth where supplies of coal could be shipped in quite easily. The windmill must have become less competitive after the abolition of the tax on coal carried by sea in 1831, which was one factor helping to introduce steam drainage into the Fens. In corn-growing areas with a reasonable wind regime, the windmill remained competitive even against steam until the closing years of the nineteenth century. It could still provide a service by grinding locally produced grain for a local market. However, this position was changed by the opening up of world markets for grain and also the demand for finer, purer flour.

The steam engine may have had a greater effect on the decline of windmills, not so much through being a direct competitor as a source of power for grinding but through its application to transport. While railways could carry flour or grain more cheaply from the countryside to the large cities in Britain, they had a much greater impact in North America where the Great Plains and Prairies were opened up. The repeal of the Corn Laws, which was completed in 1849, meant that grain could be brought freely into Britain and cheap hard wheat from America, which was the best for making the white bread so favoured by the English, began to come flooding in after about 1870. The grain was brought to American ports by steam railway locomotives or by steam river boats. It was shipped across the Atlantic by another application of the steam engine, the compound condensing marine engine which, after its development in the late 1850s, made trans-oceanic merchant ships competitive with sailing vessels. It was unloaded in British ports where stationary steam engines, themselves economical compound condensing types, ground the wheat into high quality white flour.

Agriculture in England was adversely affected by the introduction of Free

Trade in 1875 and this was made worse by a series of bad harvests between 1874 and 1882, particularly in 1875, 1876 and 1877. The peak year for wheat production in Norfolk, Suffolk and Essex was in 1874 and between 1868 and 1901, acreage of wheat decreased in those counties by thirty-two, thirty and thirty-nine per cent, respectively. The words of Longfellow's poem *Song of the Windmill* were true no longer:

> I look down over the farms;
> In the fields of grain I see
> The harvest that is to be;
> And I fling to the air my arms,
> For I know it is all for me.[3]

Regional factors conditioned the decline in windmills, for in Somerset it had set in earlier and shows again the close links between production of corn and the windmill. Here the peak year for wheat was 1868 and by 1901 the loss of arable acreage was fifty-six per cent. Decline in windmill numbers had started as early as 1840 because improved rail links made possible the transport of milk to London. Most of the region was better suited to pasture through the greater rainfall; steam ploughing or traction engines, which could till the large flat fields of East Anglia, could not be used on the small hilly ones of the West Country. Possibly because windmilling was already marginal, there were no attempts to modernise Somerset mills with fantails or patent sails and neither was roller plant introduced. While some larger watermills on rivers with regular flows were re-equipped and in Somerset may have competed with some windmills, a mere seven windmills lingered on in this county after 1900, four of which relied at least partly on auxiliary engines for power in times of calm.[4]

Then there was competition in another form, the import of flour. Figure 94 shows the impact of both the repeal of the Corn Laws and the introduction of Free Trade through the sudden rise in imports.[5] The competition from this volume of flour alone would have adversely affected the windmilling trade but there was an additional factor to be taken into consideration, the quality of the flour. Among those sending surplus flour into Britain, Austro-Hungary held a leading position. It was here that the roller mill was first developed in 1829 and the Hungarian millers retained their competitive edge by supplying flour greatly superior to anything obtainable elsewhere. In about 1870, F. Wegmann of Zurich sent the first roller mill to Britain[6] and in 1878 Henry Simon built the first complete roller flour milling plant here.[7]

The common method used in Britain was *low* milling, where grain was reduced to meal in a single pass through the millstones set close together. After this, it was put through a dressing machine to obtain fine flour. Elsewhere in Europe by the beginning of the nineteenth century the *high* milling system was used. The grain was first broken by being passed between millstones set a little apart so that, although some flour was produced, most of the grain was reduced to small particles called 'semolina'. This was cleaned to remove the fine flour and bran and then was put a second time through stones set closer together. After further sifting, the yield of fine white flour was nearly three times more than by the British method and of purer quality.[8]

Soon other machines were introduced to clean the wheat before milling. The rotary cleaner for grain, similar to a flour dresser, became common during the nineteenth century and this was followed by the 'smutter', a device with a vertical perforated drum to remove the very small spores of the fungus 'smut'. Seeds of wild corn and weeds could be removed in 'cockle cylinders', stones and grit taken out in 'rubble reels' and, by the end of the nineteenth century, the grain would be washed and dried before milling. The grinding process was slightly improved by the addition of pockets around the edge of the runner stones in which lead weights could be placed to give better balance. Then there

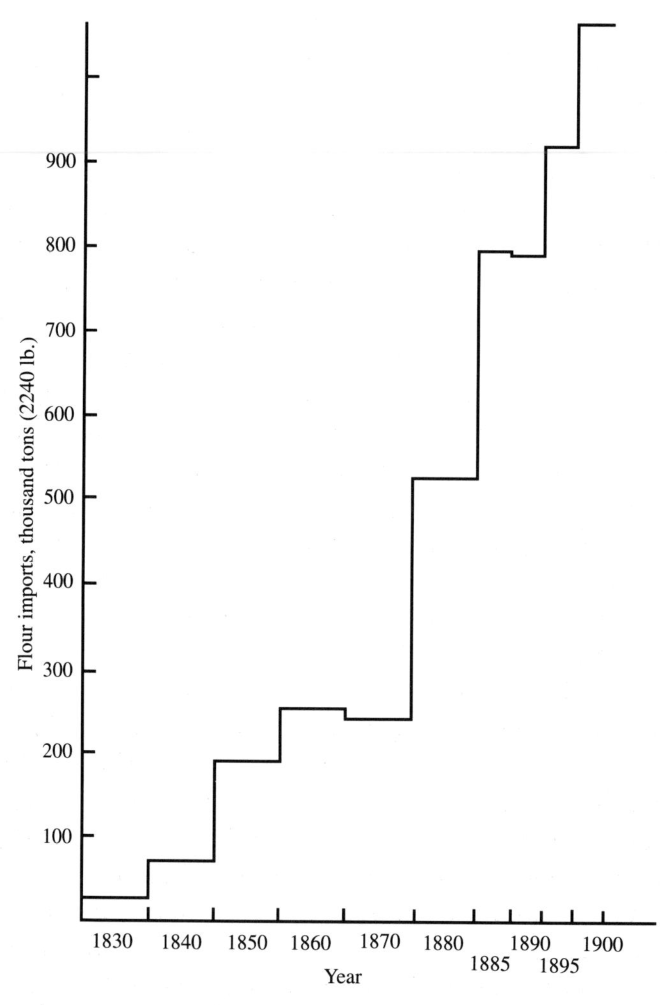

Figure 94. Imports of flour into the United Kingdom between 1830 and 1900.

were yet more machines to remove the middlings and grade the flour into more distinct qualities. During the eighteenth century, the bolter was improved as the wire machine or 'dresser'. Here a series of rotating brushes swept the meal along a stationary cylinder lined with wire gauze with graded mesh. There was another machine with reciprocating screens called the 'jumper' or 'jog-scry' so that three or four grades of flour could be produced. Even where the windmiller had space in some of the larger windmills for all this new machinery, he did not have the power because the size of the sails was limited. More and more the windmiller turned to producing agricultural foodstuffs for cattle, pigs and chickens, but even this was a declining market through competition from the larger industrial mills. The windmiller's fate was finally sealed because he could not guarantee delivery. Lack of wind might prevent him supplying flour to his local baker who, after the First World War, could telephone his nearest steam mill where the flour would be put on the petrol lorry and delivered almost immediately. The windmiller could not offer such service.

Even in the Netherlands, with its stronger winds, the rural windmill followed the same pattern of decline as in Britain but at a later date. On what were islands in the delta area of the southwest, most of the villages still proudly preserve their windmills which range from post mills to nineteenth century tower mills. One such mill at Sint Maartensdijk rivals later Lincolnshire brick mills in the elegance of its design and height (Figure 95). On the poorer lands of east North Brabant, again many villages retain their mills. It was thought worthwhile in 1908 to move one of Honig's redundant papermaking smock windmills, De Bijenkorf built originally in 1695, to begin a new life grinding corn in the village of Gemert near Helmond. Nearby at Wanroij the tall three-storeyed post mill, De Ster, was modernised with streamlined Bussel sails in the late 1930s and so this type of milling remained viable in these remoter areas. Industrial mills had to compete against the steam engine much sooner. In the Noorderkwartier, there were just under two hundred industrial mills in 1630, rising to a peak of around 670 in 1731. However, numbers had declined by about one hundred in 1795 so there was not the dramatic sustained industrial expansion which happened in Britain.[9]

During the years just before 1750, Dutch prosperity vanished. Shipping and fishing industries practically ceased to exist. Population declined by 40 per cent. There were three main reasons for this. First, repeated outbreaks of cattle plague together with a fall in the price of cheese and meat reduced the profitability of agriculture and so the basis for taxation was reduced. Then, in 1730 appeared the ship-worm, whose activities ate up the wooden sea defences. The only remedy was to build in stone instead of wood.[10] This involved great expense because so many structures had to be rebuilt. The third major problem was the silting of the river beds. Floods could no longer deposit silt onto the land and, because there was insufficient speed in the flow to carry the silt out to sea, it settled in the beds of the rivers with the result that they became shallower. Not only did this increase the risk of inundation because the floods could not escape quickly, but it affected shipping.

In 1650, Holland had 14,000 to 15,000 ships, about 60 per cent of the European trading fleet. They were larger and faster than those of 1600 as well as those of their rivals, but then the size could be increased no more owing to

Figure 95. The tall tower mill built in 1868 at Sint Maartensdijk, Tholen Island, Netherlands (August 1990).

the lack of depth in the rivers, so by 1700 French and English ships were bigger and deeper. In about 1750, the depth below water level of a British ship was 18½ Amsterdam ft. and of a French ship 20¾ ft., but a Dutch one only 16 ft.[11] The Dutch ships had to be built broad and shallow, so they were clumsy and slow, which affected trade. The Dutch navy could not protect its nation's interests across the world with these small ships. Deeper ships meant more carrying capacity, higher speed and so a much higher earning power. The French, and more particularly the British, were able to take advantage of their better harbours and capture the trade from the Dutch. The solution to river siltation was not found until the steam dredger was developed after 1860.

After 1760, the general situation began to improve, partly through tax facilities granted by the States of Holland, but there was no question of the return of former prosperity.[12] The Napoleonic Wars and the naval blockades during them wrought havoc on the Dutch economy. The separation of the Low Countries in 1830 into Belgium and the Netherlands brought further economic readjustment. In the meantime, the Industrial Revolution in Britain had changed the course of world history. It was based on the mineral resources of coal and iron and the introduction of the power of steam. While the Belgians were able to follow the British lead with their reserves of these minerals, the Netherlands had no iron ore and deposits of coal were found in the Limburg area only in the late 1840s. Windmills were ill-suited to power the new industries of the Industrial Revolution, which at first used waterpower during their early years of development, but it was the steam engine which allowed growth to continue and the steam engine required coal. Of course alternative fuels such as peat or wood could have been and were burnt, but peat is bulky for its calorific value and the Dutch did not have the resources of wood for this to be a viable possibility on any scale. If the Dutch economy were to move into modern industries during the nineteenth century, the price of coal became critical. This had to be seen against the background of the deterioration of the rivers because adequate transport facilities were vital for cheap conveyance of a bulky, heavy material like coal. The cotton textile industry in Britain too was dependent upon good communications, as the digging of the Bridgewater Canal and then the construction of the Liverpool and Manchester Railway show but, in Britain, the steam-powered cotton mills were sited on top of coal seams in the Manchester and Glasgow areas.

The coastal regions of the Netherlands should have been at no disadvantage in purchasing coal when compared with London because ships from, say, Newcastle upon Tyne could have sailed as easily to both places, although allowance must be made for the state of the Dutch rivers. However, when compared with both Britain and Belgium, the Netherlands had higher export and import duties. Roentgen, founder of the Fijenoord iron works at Rotterdam, compared the costs of coal from Britain and Belgium in 1830.

Price comparison of English and Belgian coal at Rotterdam, 1829–30 (guilders per 1000 kg)[13]

		Ghent coal		Llangenneck coal Aug. 30	
	1829	Jan./Feb. 1830	Aug. 1830	English ship	Dutch ship
Pit head cost	10.91	12.86	16.00	5.70	5.70
Freight & insurance	1.90	2.10	?	6.00	6.50
Export duty	–	–	–	3.45	6.00
Total	12.81	14.96	?	15.15	18.60

Source: A.R.A. Nationale Nijverheid. Enclosure dd 27–9–1830, no. 8N/N in 30–10–1830, no. 1A.

The last months of Union were forcing up pit-head costs in Belgium to a point where British imports were becoming viable in spite of higher freight charges and the addition of export duties. In April 1831, there was an import duty of f 7 per tonne on English coal, but this was removed by the revision of the tariff that year and industry was largely exempt from later excise duties. If the position of the coastal towns was one of severe disadvantage, the situation was worse in areas inland owing to the cost of transport.[14] When the Fijenoord iron works was buying 1000 kg of coal in 1843 at f 11, the average price in the province of North Brabant ranged between f 16.50 and f 14.33. While in neighbouring Limburg it was only f 7.08 in 1844, the price in Brabant was f 14.10. Limburg was in an exceptional position, first through its proximity to Belgium, and secondly because it possessed its own mines. By 1850, not only had the price of coal fallen but the British export duty had been abolished with the adoption of free trade. The true problem for the introduction of steam power into Dutch industries can be seen from a comparison of prices in 1855 for 1000 kg of coal. In Enschede it was f 17.62, in Leiden nearer the coast, f 11.30, but in Manchester f 4.50.[15]

Up to 1861, there were only seven steam engines in the major industrial centre of the Zaan region which included the one in papermaking, two in oil mills, one corn mill and one making sailcloth. Generally the types of industry where steam engines were introduced were either those needing large amounts of power, like oil making and rice hulling, or those where a steady power was required, like iron manufacture, weaving or the papermachine. Sawmills had cheap fuel from their own waste. The following list of mills does not cover every category.

Issue of new licenses to operate a steam engine

Type of industry	To 1861	1861–70	1871–5	1876–80	1881–5	1886–90	1891–5	Total
Sailcloth	1	1				2		5
Leather	1							1
Pumping	1		1	4	1		1	8
Papermaking	1							2
Oil	2	3	4	3	1	2	1	16
Rice hulling		2	2	3	2		1	10
Corn	1	1						2
Sawmill		3	8	12	1	2		26
Starch		3		2	2			7
Iron working				3	2		1	6
Numbers in Zaan region	7	18	22	38	17	16	10	128[16]

What these figures do not show is whether these industries increased in size and outstripped the power that could be supplied by windmills. Photographs of the later steam-powered mills show much larger buildings which suggests that increasing scale of operations was one cause of the change to steam. In 1850, only 10 per cent of the power used in Dutch industry was produced by steam. In 1904, this figure had risen to 81 per cent with wind and internal combustion engines providing 11 and 8 per cent, respectively.[17]

Steam power also failed to make great inroads into Dutch land drainage before 1900. The decision on what type of pumping equipment should be employed in any polder rested with members of the administrative board which consisted of the local people, mostly farmers or landholders. Heated arguments raged and the secretary of the Nieuwkoop polder recorded that he was unable 'to take notes'. At one meeting, the minutes read, 'Also, with an eye on the diversity of interests, the Commission considered it impossible to come to a decision on steam drainage.'[18] The change to steam drainage came through a better understanding of agricultural needs because, if crops were to be sown in the spring, the water had to be cleared from the land in good time. As long as pastoral needs predominated, there was little incentive to alter traditional practices, so it was normal for parts of polders and low-lying areas to stand underwater during the winter months. Often farms and villages would be isolated from the outside world. While the windmill provided a reasonable drainage under normal conditions, it could not cope with exceptional climatic events. During July and August 1865, severe thunderstorms ravaged the Nieuwkoop polder and the surplus water could not be removed because there was insufficient wind to work the windmills. Even the steam engines on the Haarlemmermeer polder proved inadequate on this occasion. A second inundation, in the autumn of 1870, convinced the landowners in the Nieuwkoop polder of the unreliability of windmills and a meeting in January 1871 decided strongly in favour of steam engines.

The steam engine not only quickly removed excess water in extreme conditions but at other times it could maintain the water-table at any predetermined level. As in the Fens, the recognition of this point was one of the crucial factors in the change to steam. The chances of a failure in harvests became fewer and crop yields were improved if the water-tables were kept at optimum levels. The farmer could cultivate whatever crops he liked, whenever he liked, and preparing the ground and even sowing the seeds could be done in winter. Market gardening became a possibility and it is interesting to note that the nurserymen of the Low Boskoop polder who grew trees were amongst the first to change to steam. There were other factors favouring the steam engine, for it could be sited in the best situation to drain a polder without having to avoid buildings or trees. The size of storage reservoirs could be reduced, which freed land for agriculture or other uses.

Between 1850 and 1877, the prices of land and agricultural products rose continually and it was in this period that steam engines became widely used for land drainage. In 1845, there were two steam drainage systems but forty years later there were 433. Numbers increased steeply between 1870 and 1885 but then levelled off. In 1896, 41 per cent of the drained surface area in the Netherlands was kept dry by windmills and 59 per cent by steam engines. One reason for the slow change to steam was the economic factor. The operating cost per annum for windmill drainage was f 115 per horsepower. Steam drainage for small-scale engines originally cost between f 425 and f 475 in 1826 but this was reduced to f 160 in 1891.[19] Only large steam engines like those on the Haarlemmermeer, which cost f 85 per horsepower, could beat windmills in running expenses but this was not the only point that had to be taken into consideration. The move away from wind was further hastened when internal combustion engines became available, followed by electric motors. With these forms of power available, annual expenses were further reduced and these alternatives became more competitive with wind.

After the First World War, interest was aroused in the Netherlands which led to the final development of the traditional windmill but, although the performance was dramatically enhanced, it was evident that such mills were unsuitable for further improvement, in particular for generating electricity. Albert G. von Baumhauer was first to consider mill improvement through the application of aeronautical principles. He was connected with the Laboratory for Aeronautics at Amsterdam, where there was a wind tunnel. He found that engineers and millwrights of former centuries had, by experience and intuition, developed sails which left little room for improvement. He applied for a patent in November 1918 (granted in 1922, No. 7177)[20] for his design with a curving streamlined foresail which smoothed out the front edge of the stock while the driving side was shaped into a taper so that it, too, was streamlined. Part of the driving side consisted of shutters like those on English patent sails except that they were made thicker and tapered to give a smooth surface to both sides of the sail when closed. When closed, there must have been an element of low pressure drawing on the reverse side through the aerofoil shape, but this would have been destroyed once the wind opened the shutters and it would seem that the concept of the wind pushing against the sail was still paramount. Von Baumhauer wished to try a set of sails shaped like the blades of aeroplane propellers but no millwright was willing to build them. He also had realised that, from the amount of sideways thrust between the brakewheel and the wallower, the power being generated could be determined and that this could be measured by the pressure on the upper bearing of the vertical main shaft. He designed a device for this which was tested in 1926 on the Nieuwland Mill near Schiedam and found that, at a speed of 14 r.p.m., about 40 h.p. was being transmitted.[21] This principle was used in later tests.

To understand the background of the tests which followed, it is necesary to turn to 1923 when the Society for the Preservation of Mills in the Netherlands was founded. Shortly after it was established, a Technical Committee was set up to study ways of improving performance of traditional mills and included among its members some well-known engineers from mining and agriculture, and millwrights such as the Dekker brothers, as well as von Baumhauer. A competition with a prize of 500 guilders was advertised; a jury was appointed; in 1924, forty people sent in suggestions so that the original panel was enlarged. Some small prizes were awarded, among them one for an idea for attaching an electric motor which was first tried at the Benthuizer Bovenmolen in 1929.

More important were the winning improvements put forward by Adriaan J. Dekker of Leiden. He proposed using Archimedean screws and rollers on the main windshaft bearing so that mills could work in lower wind speeds but his designs for steamlining the leading sides of sails were more significant. In 1926–7, on a drainage mill to the north of Waardenburg, he fixed some sheets which were wrapped around the leading side of the stock to streamline the flat front and they continued until they joined again at the driving side to smooth that too (Figure 96). The main portion of the driving side of the sail remained covered in canvas which could be furled in strong winds. The design of these sails had been tested first in the wind tunnel at Amsterdam and the new 'Dekkerised' sails on the Waardenburg mill were tested on 14 and 15 December 1927. The results were inconclusive and it was realised that several weeks, if not months, of trials would be necessary if a clear insight into the performance of any mill were to be achieved.[22] However, the destruction by fire of the

Benthuizer Bovenmolen in 1928 and its subsequent rebuilding by Dekker in the following year presented the opportunity to modernise it with a turbine pump and Dekkerised sails. This was the mill in which an additional electric motor was installed to drive the pump in periods of calm. The tests carried out by H.A. Hoekstra in 1929 were able to relate the average velocity for the ends of the sails to the average velocity of the wind and to determine the tooth pressure, and hence the power output, more accurately as the dynamometer was read systematically.[23] A mill with a scoopwheel at Lopik was tested in May 1931 and showed that modernised sails performed better than traditional ones, but that this was nullified at the Benthuizer Bovenmolen by the relatively high losses of energy in the turbine pump owing to the irregular speed of that mill.[24] Dekker was granted patent 24753 in 1929.[25] In 1930, thirty-one mills were fitted with his design of sails, in 1933 there were fifty-six and in 1935, the total had reached seventy-five.

In January, 1935, the Catchment Board of 'Schieland' purchased the Prinsenmolen, a large drainage mill with a scoopwheel near Rotterdam originally built in 1648. The engineer in charge of the restoration and modernisation of this mill took the initiative to carry out tests on it before and after reconstruction. Its situation, in an area free from obstructions by the side of a large lake, suggested that it would be well suited for experiments because measurements of wind velocity would be more than usually accurate. A weir for measuring the water was installed on the outlet to the scoopwheel and two towers built nearby on which anemometers were mounted. With the original scoopwheel, the mill did not have any useful output when the average wind velocity was less than 6.5 m/sec. (14 m.p.h.) and that really a velocity of 8 m/sec. (18 m.p.h.) was necessary to raise any water. Improvements to the scoopwheel carried out during the winter of 1937 slightly improved the performance but still showed the limitations of traditional mills. Attempts to test this mill with a Prony brake were a failure.

The people who had been carrying out these tests, the 'Prinsenmolen' Committee, had placed at their disposal in 1938 a small mill built as a training exercise for unemployed workmen in 1935. It was moved to a site in open fields

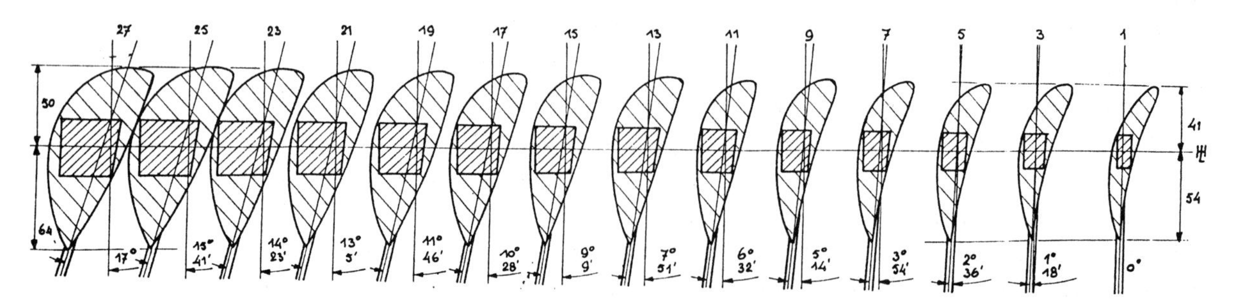

Figure 96. The layout of the sheathing over the leading edge on Dekker sails.

north of Delft and fitted with a Prony brake. Anemometers were installed nearby. At first, tests were carried out with a design of sails taken from one of the famous Dutch windmill books of 1734. In gusty wind, the mill ran very irregularly. To help compensate for this, lead weights were fitted on the sails to increase the mass and were an immediate success. In the meantime, wind-tunnel tests had been carried out on various streamlined sails. Although the principles of Sabinin and Bilau could not be followed closely in practice, these experiments showed that the irregularities of the sail surface in traditional Dutch designs caused areas of turbulence. This was one problem with the original sails on the Prisenmolen where, in 1939, new sails with streamlined stocks were fitted and proved to be of great value during the Second World War, when practically no diesel oil was available and electric power plants lacked coal or other fuel.

During tests on the Prinsenmolen in the autumn of 1939, it was established that a gust of 6.5 m/sec. (14 m.p.h.) was sufficient to start the mill and that it ceased to pump water when the velocity fell to 4 m/sec. (9 m.p.h.) for 15 seconds or more.[26] These improved windmills would start to turn with a wind speed of 3.5 to 4 m/sec. (7¾ to 9 m.p.h.) compared with 5 to 6 m/sec. (11 to 13½ m.p.h.) for the old. At 5.5 m/sec. (12½ m.p.h.), their power would be equal to that of a normal mill at 8 m/sec. (18 m.p.h.) and they would be running much faster. In Holland, there are winds with speeds between 4 and 6 m/sec. (9 to 13½ m.p.h.) for 1,771 hours per annum, between 6 and 8 m/sec. (13½ to 18 m.p.h.) for 1,332 hours and between 8 and 12 m/sec. (18 to 27 m.p.h.) for 1,339 hours. This shows that a normal mill can be worked for around 2,671 hours compared with 4,442 hours per annum for the improved mill, an enormous difference.[27] The annual output on an improved windmill may therefore be assumed to be approximately twice as great as that of a normal windmill. It was not until 1940 that the 'Prinsenmolen' Committee fitted onto their model mill fully streamlined sails resembling a propeller profile which were based on a design of Sabinin and Yurieff; it was quickly established that, with a moderate wind of average velocity, they developed about two and a half times as much power as Dutch sails.[28] Unfortunately the war intervened and stopped further investigations but already these sails had shown the direction that future developments must follow.

There were many other investigations carried out in the Netherlands, into both the performance of windmills and the wind, so that gradually a better understanding of the problems was developed. Following Dekker's lead, other millwrights experimented with their own systems. Among them was Chris van Bussel in Weert who, in 1934, modified the Dekker leading side by making the front part concave so the wind was guided better onto the sail (Figure 97). He retained the cloth-covered driving side because there was still a desire to keep the traditional appearance of windmills.[29] However, his streamlined casing did not stretch beyond the stock so there must have been considerable turbulence at the stock where the cloth started. Between 1935 and 1937, Bilau not only streamlined the stock but led this into a streamlined driving side as well. The full length of the driving side was made into a single board which could pivot to spill the wind. A similar idea by Ten Have was combined with the streamlined leading side of van Bussel (Figure 98). These designs attempted to reduce turbulence across the blades and, by their aerofoil shape, to introduce an element of low pressure behind the sail.[30] The single longitudinal shutter, instead of the many small shutters found on English patent sails, would have been stronger

Figure 97. This mill at Reek, North Brabant, has been fitted with van Bussel's imporoved leading edge (August 1990).

Figure 98. The Agneta mill at Ruurle, Gelderland, has Ten Have longitudinal shutters fitted on the driving sides of the pair of sails in the horizontal position and Fok's foresails fitted on all sails (March 1992).

and had fewer moving parts to maintain. Also, the longitudinal shutter gave a better aerodynamic performance and may even have acted as an air brake in strong winds. The final modification was P.L. Fauël's adaptation of the leading side or foresail, again around 1935. Cloth was retained on the driving sides but the foresail was made from a single board curving around the leading edge of the stock with a gap between it and the stock where the wind could pass through (Figure 99). The foresail itself must have helped not only to guide the wind around the stock and onto the cloth but also acted as a sail itself and probably gave more power to the mill. Mills at the Aarlanderveen polder, which are still the only method of draining that area, at one time had some form of Dekkerised sails but today are entirely equipped with the Fauël foresails.[31] The tips of these foresails were sometimes fitted with sections that could be rotated to act as wind brakes.

It was shown that, by careful improvements, a Dutch drainage mill could develop a reasonable horsepower with the average wind velocity as low as 6 m/sec. (9 m.p.h.). In 1947 a committee was formed to carry out more tests on the Benthuizer Bovenmolen in order to explore more deeply the possibilities of

Figure 99. Fok's foresails have been further adapted with the end portion able to rotate and act as a wind brake on this corn grinding mill at Witmarsum, Friesland (August 1990).

generating electricity. Modifications to the mill were finished on 1 October 1950 and tests continued until the following September. It was calculated that the consumption of electricity would have been 15,460 kWh higher if the electric motor alone driven had driven the pump. In addition, 4,730 kWh had been generated by the motor when driven by the sails and acting as a generator. So the wind had provided 20,000 kWh in eleven months, an amount which could have been greater if the mill had been worked on Sundays and public holidays or when the water-level in the polder did not warrant it. The mill was also out of commission for three days in October 1950 and for two months in March 1951 through the wooden gear teeth stripping. Wooden gearing proved to be unsuitable. When a windmill is equipped with an Archimedean screw or any other old-fashioned machinery, a squall of wind will cause the sails and machinery to run faster so that the tooth pressure remains constant. But when the windmill is driving an A.C. generator, the speed will remain constant and therefore the tooth pressure may become excessively high whenever a gust of wind hits the sails. One answer was automatic reefing or pivoting sails instead of traditional ones. Various endeavours were made to adapt Dutch mills to generate electricity but they all encountered the technical problem of insufficient strength in the gearing. To modernise such mills meant completely redesigning the gearing and it seemed unlikely that this could be justified.[32] However, some mills were converted to generate electricity. The last was the Traanroeier mill at Texel which, after testing, went into operation on 15 September 1965. The sails were fitted with spring-loaded aluminium shutters similar to English patent sails and braking flaps. The leading sides were built as aerofoils and had small slits in the streamlined leading edges of the steel stocks which caused a big starting torque useful during periods of light winds. With wind speeds above 11 m/sec. (25 m.p.h.), the shutters began to open to limit the power produced by the asynchronous generator.[33] The Traanroeier mill had to be stopped in 1972, partly because the rotation of its aluminium sails caused interference with television. The next generation of windmills would be built to very different designs.

CHAPTER 9

The windmill for pumping water and water supply

Britain

Early British mills for water pumping and supply

The traditional windmill with four sails gave a compromise between reasonable starting torque and speed of rotation. However, the survival of windpower required specialisation for particular tasks, as is illustrated by the cases of water supply and electric generation. Mention has been made of Ramelli's picture of a windmill raising water with a rag and chain pump to a cistern around the top of its tower, from where the water was fed to fountains in the garden.[1] There would have been few outlets for such a mill but traditional windmills with reciprocating pumps could equally well suck water out of a flooded field or quarry as pump water up to a reservoir. There must have been many cases where light wooden or steel towers supported the sails of a windmill which turned a crankshaft to operate the pump through a long connecting rod. Sometimes a more permanant structure might be provided in the form of a stone or brick tower. The remains of tower wind pumps survive in Scotland, most notably one at St Monace, Fife, for pumping sea-water into salt pans.[2] At other salt pans around the British coast, windmills could be found serving the same purpose, for such places were exposed to the winds. At Southwold in Suffolk, there was once a very primitive skeleton hollow post mill for supplying those salt pans.[3] In Kent, there were windmills beside salt pans on the Isle of Grain. At Stonar near Richborough, a picture of 1787 showed a couple of mills not marked later on the first edition Ordnance Survey maps.[4]

Late nineteenth century windmills with reciprocating pumps could be found in brick and tile works around the Humber estuary in both Yorkshire and Lincolnshire.[5] At Howden brickworks near Hull, the brick tower complete with pumping equipment has been preserved. The sail stocks were bolted to an iron cross on the windshaft which was itself the crankshaft for operating the pump. The sails were kept facing the wind by a vane mounted on a single timber beam. There were similar mills at Elvington, Yorkshire, Sutton on Sea, Lincolnshire and in the North Riding at Claxton. They would have been quite simple to construct and no doubt kept the workings clear of water in places where there was sufficient wind. In Scotland, drainage of flagstone quarries was the last industrial use to which windpower was applied to any extent.[6]

But it was mills for pumping water for domestic use in towns or, more particularly, on farms which were to have the greatest impact. In London, there must have been some form of water supply system before 1618 because Busino, a member of the Venetian embassy, commented:

> They are very badly off for water, although they have an immense supply. They raise it artificially from the stream even by windmills, and force it into all the fountains throughout the suburbs, but it is so hard, turbid and stinking that the odour remains even in clean linen.[7]

York Buildings on the Strand were the scene for many years of a water supply system. In 1673, permission was sought to change the source of supply from springs to the Thames. Christopher Wren, as Surveyor General, viewed the site and a ninety-nine year lease was granted. A windmill must have been installed for, after a fire in 1684, it came to light that the equipment was 'not built answerable to the engine first proposed by the said undertakers and approved by his Majesty's surveyor'. The King ordered that the new waterworks incorporate neither a chain pump nor a windmill, whose clatter had infuriated the residents.[8] Thomas Savery installed one of his patent steam pumps here soon after 1700. In 1725, Daniel Defoe described another windmill in London. To supply the expanding suburbs, the New River Company found that it had to build a higher reservoir at Islington which at first was filled by a wind pump with six sails.[9] This was not very successful, for London is a poor area for wind, and the mill was soon replaced by teams of horses. Also in 1725, King's Lynn was supplied with water from the River Gaywood by Kettle Mill, a smock mill.[10] In 1734, Anthony Parsons patented a form of horizontal windmill which he claimed would be 'very useful for supplying of towns, seats, houses and gardens with water' as well as mines. The water would be forced up two lead pipes to a height of 183 m (200 yards).[11]

A six-sailed windmill was installed in 1790 at Thames Head in Gloucestershire to supply the Thames and Severn Canal with water to keep the top level filled. A borehole 16.47 m (54 ft. 6 in.) deep had been sunk down to a reliable water supply and a pump with 27.9 cm (11 in.) bore installed. Samuel Ireland's sketch drawn in the summer of 1790 shows a tall thin column supporting common sails, each of which must have been about 3.05 m (ten ft.) long. He remarked that the mill was capable of throwing up several tons of water every minute.[12] Its success was its own undoing, for it showed that there was an adequate water supply and so was quickly replaced by a Boulton and Watt steam-pumping engine.

A present-day visitor to England, driving from Dover along the M2 towards London, may notice the tall tower of an old windmill near Faversham. In a way it anticipated modern practice for it was built around 1860 by the Corporation as an auxiliary to the steam-pumping engines for water supply. It developed 15 h.p. and operated a three throw crankshaft for driving the pumps which could raise 10,000 gallons an hour.[13] A powerful mill, originally with five sails, was built in 1874 behind Margate for a similar purpose. Following a severe gale in January 1878, repairs and alterations costing £250 resulted in a four-sweeps but, after being tail-winded in a high wind in August 1894, repairs costing £275 were considered too high and it was abandoned. A third mill, in Chatham, was converted from corn grinding to water pumping during the last quarter of the nineteenth century. From a well 60.96 m (200 ft.) deep, water was supplied to around a hundred neighbouring houses which brought the miller an income of about £100 per annum.[14] This was a private venture unlike the other two. Scattered about the country could be found windmills raising water, such as the one at Haigh near Wigan which supplied a brewery.[15] The little preserved mill at Starston near Harleston in Norfolk, with a hollow cast iron post, is a surprising combination of four patent shuttered sails with twin winding vanes and is about 6.1 m (20 ft.) tall. It was built by Whitmore and Binyon in about 1910 and supplied water to a farm and cottages. There were others similar in the neighbourhood.

Windmills were well suited for supplying water because, provided there was a large reservoir to act as a buffer between supply and demand, they could work whenever there was sufficient wind and store the water until needed. Neither constant power nor power in a definite amount were necessary as the pump could run with any wind of sufficient velocity to overcome the starting friction and work up to any speed within the capacity of the machinery to avoid damage.[16] Plunger pumps normally work quite slowly. Such windmills required a good starting torque to begin pumping against the static head of water and this would have been provided by the six-sailed mills in London and at Thames Head and the five-sailed mill at Margate. An increase in the number of sails will increase the starting torque but limit the speed. Therefore multi-sailed windmills are well suited for water supply.

America

The American windmill

The explorers who left Western Europe during the sixteenth and seventeenth centuries took their industrial techniques with them. This included windmills which were erected in suitable places such as those built by the Dutch in South Africa or others put up in the West Indies to drive the rollers for crushing sugar cane. But it was the windmills in North America which were to become most important for the development of windpower. When the first white settlers arrived on the eastern seaboard of the country which would become the United States of America, they found a landscape not too dissimilar from the Europe they had left. The hills were covered with fine trees and there were many rivers running down to the sea. Therefore they were able to transfer their traditional skills, which included millwrighting in wood, and establish a way of life closely based on the one they had followed back at home. The settlers were quick to utilise the abundant waterpower both for grinding their corn and for sawing their wood. It was estimated that there were some 55,000 grist and sawmills at their peak around 1840[1] as well as the famous New England textile mills.

In places where streams were lacking, the settlers erected windmills to grind their grist. Probably the first was at the Flowerdew Hundred Plantation on the James River not far from Jamestown, Virginia, in 1621.[2] Further north, the French began to build windmills on the banks of the St Lawrence River in 1629, followed by many others on Lake Erie and Lake Huron.[3] Long Island, on which New York is situated, lacked streams for waterpower and the first governor of New Amsterdam, as the city was then called showing its Dutch origins, is credited with putting up three windmills. Old drawings of New York show many windmills and it is said that ship's captains watched the windmills on the waterfront to see how to set their sails when entering or leaving harbour. Also, a law was passed in 1680 which forbade anyone to row across the river from New York to Brooklyn when the windmills were stopped through high winds.[4] While the mills often reflected in the details of their structures the European origins of the millwrights who built them, the abundance of wood meant that most were either post or smock mills covered with shingles. Construction continued until the early years of the nineteenth century. The Beebe Mill at Bridgehampton, built in 1820, was fitted with a fantail showing that American millwrights kept abreast of practice elsewhere.[5] Windmills were spread along the eastern seaboard with examples in Rhode Island, Massachusetts, Cape Cod, and even one on an artificial island in Philadelphia harbour.

In the 1860s various factors, including the ending of the American Civil war, encouraged migration westwards. Conditions dramatically changed once the settlers crossed the great Mississippi River for they entered a new world of open plains, with no trees and few streams. They had to adapt to survive. In 1862, the Homestead Law was passed which granted 160 acres of unoccupied land in the public domain to anyone who would cultivate it for five consecutive years. Although it was open to abuse by speculators, fifty million acres were distributed to prospective settlers in the next twenty years.[6] Farming in these regions would not have progressed beyond subsistence level without railways following hard on the heels of the new settlers to take their cattle and grain to the industrial towns and cities on the eastern seaboard. In 1931, Walter Prescott Webb published his book *The Great Plains* in which he put forward his thesis that the white man's conquest of this vast, semi-arid treeless region was won by the revolver which defeated the Red Indians, barbed wire which controlled the wandering of both Indians and cattle, and the windpump which provided water, all supported by the railway.[7] Samuel Colt patented his six-chambered revolver in 1838. In 1844, fifteen Rangers armed with revolvers decisively defeated seventy Comanches, the first time this had happened.[8] Although Colt's first factory went bankrupt, he was able to restart in America and, on 1 January 1853, opened another revolver factory on Millbank in London.[9] At his Hertford factory, there were 400 different machine tools because each revolver needed 357 distinct machining operations. The parts of individual revolvers were claimed to be interchangeable through this system of mass production which originated in American armouries. Colt said 'There is nothing that cannot be produced by machinery' and that his factory was 'reduced to an almost perfect system'. This 'American' system of manufacture was soon applied to agricultural implements such as reapers and also to new water-pumping windmills invented by Americans, ousting old millwrighting traditions.

Out on the treeless Great Plains, the methods of farming used in the eastern states, where cattle could be left in fields surrounded by wooden fences, had to be modified because wood was too scarce and expensive. In the early days of settlement, the cattle might be allowed to roam because there was plenty of space. The land had no value, the grass was free and the water belonged to whoever came first. But the Homestead Law began to change all this and an increasing population meant that territorial rights had to be delineated. In 1874, the first piece of barbed wire was sold in the United States.[10] Here was the answer to the need for cheap fencing, which would control the free range of the Red Indians so that the farmer could protect his homestead and install a windmill to supply drinking water; barbed wire would also control the wanderings of the farmer's cattle so that they could be protected more easily and better stock introduced; and finally, barbed wire enabled the farmer to divide his land into plots where he could install more windmills to supply drinking water for his stock and move his cattle more easily from section to section to regulate the supply of grass. He now had a system of pastures: summer pastures, winter pastures, pastures for high grade stock or for horses. Along the fertile river valleys, fields were opened up on which hay and other crops might be grown to supplement the grass out on the range.

As the new settlers moved westwards, they found that the rainfall declined. Along the southern part of the Mississippi, there is between 1,015 and 1,524 mm (40 to 60 in.) annual precipitation but further north, by the Great Lakes, this has dropped to below 1,016 mm (40 in.). In the middle of the Prairies, there is less than 50.8 cm (20 in.) annually and it is generally agreed that, with only this amount, the climate is inadequate for Western European agricultural methods. However, in the Great Plains, the summers are wet and the winters dry and it is this that has enabled the region to become a great corn-growing area. To the west of the Great Plains lie the Rocky Mountains and much of the moisture which falls on them as either rain or snow must find its way eastward to the Mississippi. Mountain streams have deposited silt across the Great Plains, through which water percolates and is trapped above an impervious sheet of marine rock inclining towards the Mississippi. Therefore by digging or drilling through the silt, water can generally be found not too far underground. However, this is available only in limited quantities because it first has to percolate through the soil. Pumps were needed which could raise the water as fast as it became available, at all hours. They had to be cheap both to construct and to operate and also capable of making a slow but constant delivery in order to raise as much as possible of the precious fluid when it was there.

At first, wells were dug by hand and might be from 2.44 to 9.14 m (8 to 30 ft.) deep, lined with brick or stone. The water would be drawn up by buckets. Such methods, both in making the well and in raising the water, were far too laborious and costly for watering large herds of cattle. Well drilling on a large scale began in the 1860s. The holes would be 15.2 cm (6 in.) in diameter and 9.14 to 91.44 m (30 to 300 ft.) deep, with a sheet-iron casing.[11] These boreholes needed a mechanical method of raising the water, and the answer was found in the windmill. The potential was enormous for the stockmen had found that cattle should not walk more than 4.02 km (2½ miles) to water. On big ranches, water was provided by regularly spaced wells, windmills and storage tanks. The famous XIT ranch, with three million acres, had over five hundred windmills[12] and there were others of similar proportions. The cowboy's duties changed: he now had to mend fences and to grease windmills.

Compared with the humid East, the Great Plains is a region of high wind velocity.

> 'Does the wind blow this way all the time?' asked the visitor to a ranch in the West.
>
> 'No Mister', answered the cowboy; 'it'll maybe blow this way for a week or ten days, and then it'll take a change and blow like hell for a while.'[13]

Over this treeless open country, there was little to retard and restrict the winds so they blew with remarkable uniformity and with a relatively high speed, averaging 4.75 to 5.36 m/sec. (10 to 12 m.p.h.) and even reaching 6.25 to nearly 7.0 m/sec. (14 to 15 m.p.h.). But such winds would not have been very effective with traditional windmills.

Some traditional windmills were, however, built in the West for grinding corn or grist. There was Robert Logan's mill, begun in 1825 at Fort Douglas on the the Red River in Manitoba, and another at Colorado Springs which continued to grind until that town was reached by the Denver and Rio Grande Railway in the 1870s.[14] The wooden construction of such smock mills covered with boarding would have made them very expensive for the average farmer of the small homestead, even if these mills had been suitable for pumping water in such comparatively light winds. What was needed was a mill which would have good low-speed torque for starting, be able to keep on running in relatively light winds, face those winds from whatever direction they came, and cease to operate if the wind became too strong. All this had to be done with the minimum of attention, for the rancher had to be away looking after his stock and had time only to give the working parts of the mill an occasional greasing. Therefore the Americans had to evolve a self-regulating windmill which could be left to fend for itself and operate even in the worst of their winters.

Part of the solution was found in the annular sail, winded by a vane. First, the number of blades round the circumference of an annular sail presented a large surface area on which the wind could press and give enough torque to start a piston pump working. Such pumps did not need to operate very quickly and this matched the type of rotor. The speed of the rotor is linked to its 'solidity', defined as the percentage of the circumference of the rotor filled by the blades. The greater the solidity of a rotor, the slower it turns and hence the greater the low-speed torque. Most of the early American disc windmills were built with blades made from flat wooden strips and their total area was equal to the swept surface. If all the blades were placed flat and side by side, they would form a complete circle. Some makers went further and, under a mistaken conception, enlarged the sail surface until it exceeded the area of the annulus but this was later reduced to about 66 per cent.[15] The typical solidity of a modern multi-bladed rotor for water pumping is 40 to 60 per cent.[16] On many of the early mills, the blades were straight with a weathering of about 35 degrees which also helped to give a good starting torque. Another advantage of both the solidity of these wheels and the weathering at the tips was that the speed of the wheel regulated itself because the blades acted as a kind of air brake when abnormal speeds were approached. These mills generally started turning with a wind speed of about 2.24 m/sec. (5 m.p.h.) and some form of governing system had to be introduced when the wind reached about 7.15 m/sec. (16 m.p.h.).[17] The pump load increased directly as the speed of rotation but the wind pressure varied with the cube of the wind speed, resulting in an excess of power as the wind velocity rose. Reducing the power output could be done by twisting and feathering the blades, by tilting them or by turning the whole rotor out of the wind. The development of the American windmill covered many different ideas as hundreds and thousands of them were built in the last half of the nineteenth century.

The first small windmills suitable for use on homesteads appeared in the 1850s. On the Pacific coast, William Isaac Tustin stated that, in 1849, he erected the first windmill in California at Bernica to supply water for the community. It is not clear when he began serious manufacture, for the first definite reference to windmill production in San Francisco was not until 1868. He sold his factory and rights in 1879.[18] Another early pioneer was a missionary among the Shawnee Indians, Allen T. Ward, who wrote to his sister in September 1850 that his latest experimental windmill was very satisfactory and 'proved that it was the proper power for this country'.[19] It was on the self-adjusting principle with springs which allowed the sails to yield to the force of the wind. He did not patent his mill but, during the 1850s, over fifty windmill patents were submitted to the U.S. Patent Office.

In the 1840s, John Burnham, who was born at Brattleboro in Vermont, had considered the idea of harnessing the power of the wind, but he lacked practical expertise. His business was manufacturing hydraulic rams for domestic water supply, which were distributed throughout New England and elsewhere. In the 1850s he returned to the idea of windpower for he saw that there was a potential market in areas without sufficient streams to operate his hydraulic rams. It was probably in 1853 when he approached Daniel Halladay about designing a windmill which would run without any attention from a person either to turn it into the wind or to control the speed and thus prevent it destroying itself in high winds. Halladay had been apprenticed as a machinist and, when Burnham approached him, he owned machine shops at Ellington, Connecticut. Within a remarkably short time, Halladay produced a self-regulating windmill. Halladay's mill had a fixed vane which kept the sails winded (Figure 100). There were four solid sails or blades which were mounted on pivoting radial rods in a central casting. In light winds, the blades faced the wind at the normal angle of inclination but, as the wind increased and the mill ran faster, a centrifugal governor changed the pitch through a linkage system. When the winds were high, or when the mill was switched off by the operator, the blades were turned into a position parallel with the vane, and also the wind, which stopped the mill moving. The whole mechanism was mounted at the top of a tall strong post and the pump in the well was operated by a shaft connected to a crank formed on the windshaft itself. In September 1854, the first mill had been under test for six months pumping water out of a 8.53 m (28 ft.) deep well and raising it to a height of 30.48 m (100 ft.). More than sufficient water had been provided

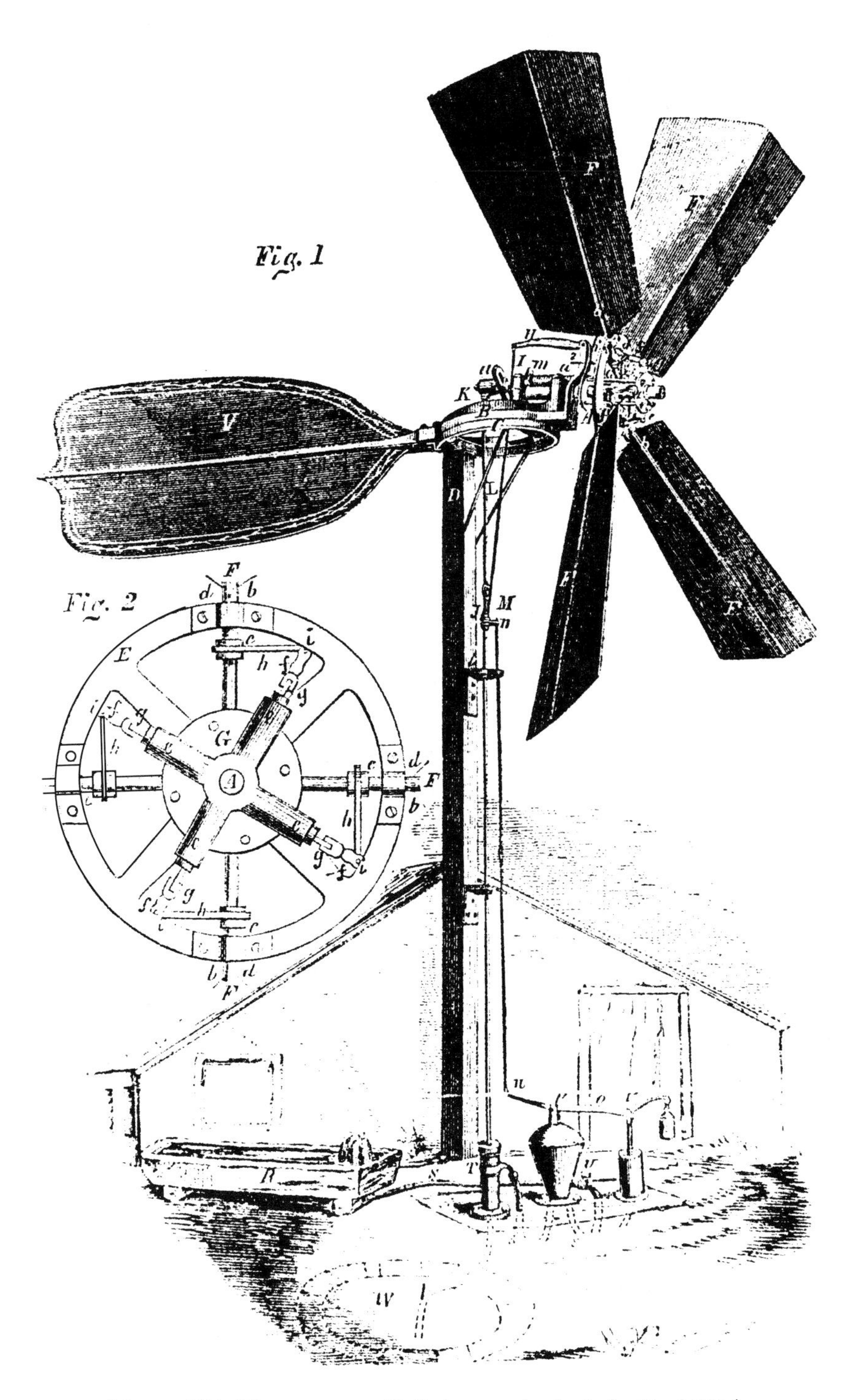

Figure 100. The prototype Halladay standard windmill of 1854.

for farming purposes and garden irrigation. The mill itself had needed no regulation of the sails and it had run fifteen days at one time without stopping day or night.[20] A patent was taken out in August 1854 and manufacture commenced at Ellington but was later moved to South Coventry, Connecticut. One mill was exhibited at the New York State Fair and was awarded the highest premium. However, sales in New England never achieved the volume its promoters desired and, in 1856, John Burnham moved to Chicago because he realised that the Midwest and its thousands of farms without running water presented a potential market. A year later he organised the United States Wind Engine and Pump Company as an agency for distributing the products made by the Halladay Wind Mill Company back in New England. Partly owing to the constraints of transport during the American Civil War, the U.S. Wind Engine Company purchased the Halladay Wind Mill Company and shifted the entire manufacturing operation to Batavia, Illinois, about 56.33 km (35 miles) west of Chicago, where it remained until it closed at the start of the Second World War.

In 1855, John Burnham found another outlet for his windpumps when he sold one to the Illinois Railroad Company of Chicago, which used it to supply water for locomotive boilers. Such windmills became prominent features of railway stations as the railways spread across the plains and prairies into the West, to supply water to thirsty people and steam locomotives alike. The early Halladay mills in the 1850s were based on a design with six or eight radial blades ranging in size from 1.83 to 4.88 m (6 to 16 ft.) in diameter. The smaller mills had solid wooden blades while the larger models used blades made of iron frames covered with sailcloth.[21] Soon after production was moved to Illinois, the style of the rotor was changed and remained virtually the same for seventy years, creating one form of windpump which would become common throughout the West. The new mills had appeared by 1868, which is after annular sails had been offered for sale by competitors. Many thin wooden blades were mounted in six sections on straight pivoting slats which formed chords around the circumference of the rotor (Figure 101). The pivots of the chords were bolted onto radial arms and each section had a radial operating rod which controlled how far the chords and their connected blades could pivot. As wind speed increased and with it the revolutions of the rotor, centrifugal force pushed out the ends of the blades closest to the centre and made the whole section tilt away from the wind, producing a larger opening in the middle through which the wind could pass more easily. The centrifugal force was countered by a series of rods linked to adjustable weights to give the required speed. These weights closed the sections when the wind dropped so that the whole action could be pre-set and was automatic when running. A fixed vane winded the rotor. These sectional tilting wheels never achieved the popularity in England that they did in America. One reason may have been that the overhang of the rotor had to be quite large to avoid the sections touching the tower. Another may have been

Figure 101. A mill with tilting vanes manufactured by the Continental Wind Mill Company of New York, U.S.A., in the 1860s and 1870s.

that, to English minds, the appearance was unsightly, being suggestive of a damaged umbrella in strong winds.[22]

Just before the turn of the century, the U.S. Wind Engine and Pump Company was making mills in 'farm sizes' of 3.05, 3.66, 3.96 and 4.27 m (10, 12, 13 and 14 ft.) diameter. There were 'railroad sizes' for pumping from deep wells or raising a larger volume of water from shallower wells in 4.27, 4.88, 5.49, 6.1, 6.71, 7.62, 8.53 and 9.14 m (14, 16, 18, 20, 22, 25, 28 and 30 ft.) sizes. In addition to the pumping mills, the crankshaft could be replaced by a bevel drive and a rotating shaft down the tower for driving machinery. These had appeared as early as 1855 and some were made in sizes up to 18.29 m (60 ft.) in diameter for powering grist mills and factory equipment.[23] The problem here was how to keep the rotor facing into the wind, as the torque exerted through the drive shaft would try to rotate the whole head; the fantail provided one answer.

Two of the predecessors of Halladay and Burnham with the annular sail were Frank G. Johnson and Francis Peabody. Johnson, of Brooklyn, New York, was securing patents as early as 1855 and in the following year was selling mills as far away as Kansas. A drawing of his mill shows that it was winded by a vane (Figure 102). There were nine sails in a ring, mounted in groups of three. The governing system operated rods which twisted the sails to spill the wind.[24] These mills remained on the market at least during the 1850s. Mills with twisting blades never achieved the popularity in America that they did in England. Possibly the more severe winter weather conditions in the States iced up the many working parts. Such mills were less likely to be damaged in storms because the blades could be turned edgewise and presented merely a skeleton to the wind. On the other hand, they may have been less sensitive to governing because the blades had to open in the direction the wheel was revolving and so against the resistance of the air.[25]

Francis Peabody, of Salem, Massachusetts, has been claimed to be the first inventor actually to produce mills with annular sails of sufficient size to operate flour and grist mills, for in 1857 he erected an experimental mill in Illinois to test his theories and patterns.[26] This is of course later than the large annular sails made by Chopping in England. Francis patented his ideas in England under the name of Joseph Peabody. He used a vane to wind his mill, which he described with a pump rod descending through the middle. He stated that

> The force of the wind is graduated by a disc of wood or sheet metal secured to the shaft. The wind wheel has the inclined vanes permanently secured to the rim and the hub and is fitted to the shaft by a feather, so as to slide along the shaft. It is usually kept some distance in front of the disc by a helical spring, but when the force of the wind increases, the wheel is made to approach the disc, and the passage for the air between the two is contracted, and the velocity of the wheel is thus checked.[27]

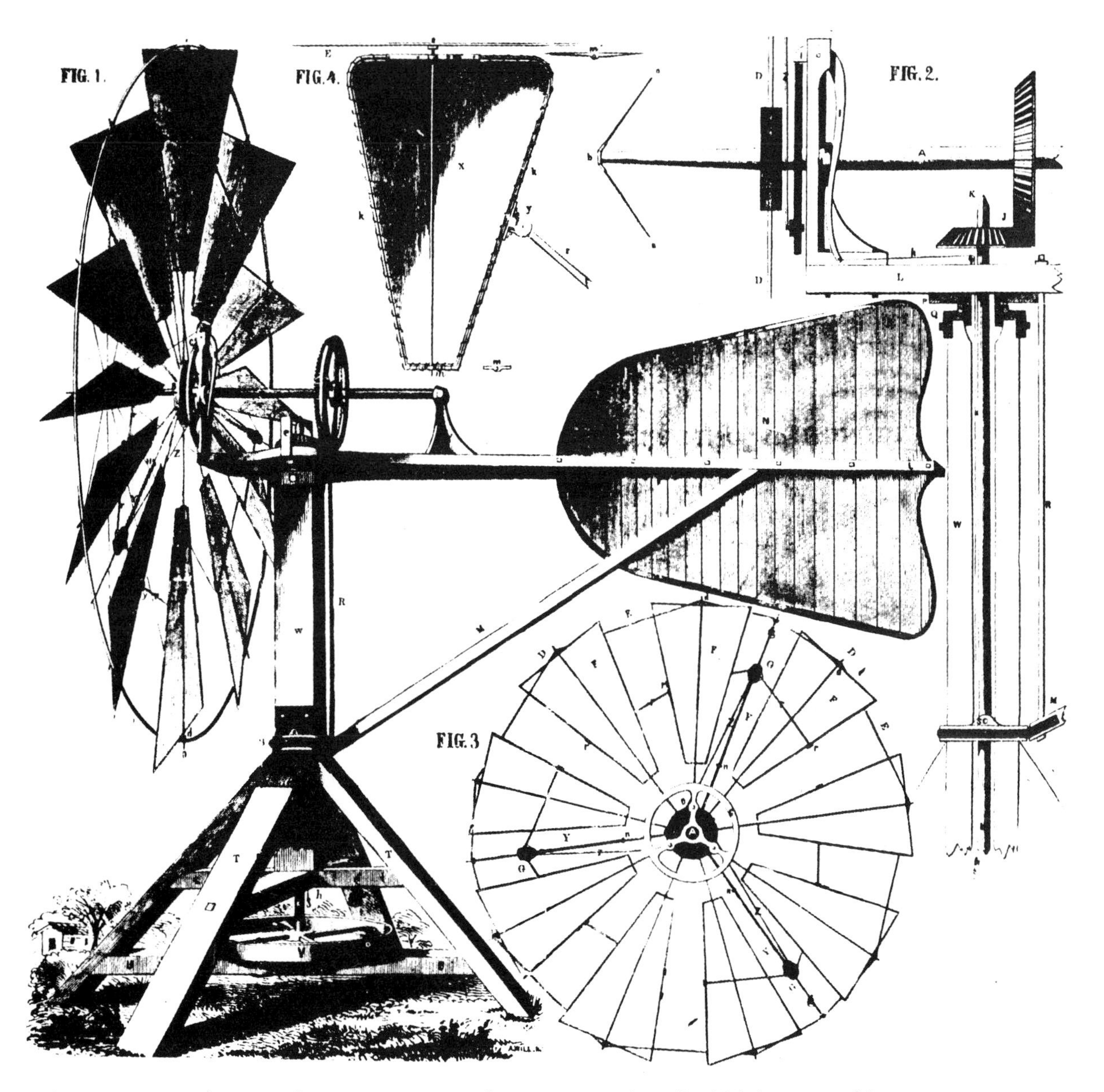

Figure 102. The drawing of F.G. Johnson's windmill with annular sail which he patented in 1855.

Nobody else seems to have used this method of governing. Peabody described another system with twisting blades but probably he built few mills.

Of much greater importance were the mills of Leonard H. Wheeler. He too was a missionary who worked among the Ojibway Indians at Adana, Wisconsin. As early as 1844 Reverend Wheeler had thought about the possibility of employing the wind to grind corn and pump water at his mission station but it was not until 1866, when he was forced to stop his normal work through a broken wrist, that he began actual construction of a mill. His son had broken a leg at the same time and the invalid pair received help with the iron parts from a nearby government blacksmith. After a storm destroyed their first mill, in 1867 the Wheelers set about building one that would be self-regulating and patented it in the following year.[28] Their new mill had four paddle-shaped blades to turn the windshaft, which was in the form of a crank for operating the pump. The blades were fixed and so, to reduce the surface area exposed to the wind, the whole rotor was allowed to turn out of the wind. This was achieved by what would become the most common method to control these windpumps. The vane winding the sails was hinged so that it could move one way to a position parallel with the rotor. It was normally kept in the position pointing away from the rotor by a weight or a spring which would cause it to return to this position when deflected. The Wheelers fitted a second smaller fixed vane parallel to the rotor protruding beyond the ends of the sails (Figure 103). The increasing strength of the wind pushed against this second vane and began to turn the rotor out of the wind but this would turn the large vane into the wind until equilibrium was struck between the two. If the wind increased further, the smaller vane would turn the mill more so that the pressure increased on the larger vane until it overcame the resistance of the spring or weight, which allowed the larger vane to pivot out of the wind; this in turn allowed the rotor to pivot out of the wind too and so reduce speed. When the wind decreased, the spring or weight returned the large vane to its proper position which brought the rotor back into the wind again, a simple solution which regulated a mill with fixed blades smoothly. To stop the mill, the spring tension would be released or the weight raised, which allowed the vane to pivot and the sails to turn out of the wind. Sometimes a brake would be applied at the same time. This type of regulation assumes a steady output or load. While satisfactory for pumping water, it does not work so well when driving machinery or generating electricity, where the load may vary.

The firm of L.H. Wheeler and Son began manufacturing their mill at Beloit, Wisconsin, in 1867 and two years later the paddle blades were changed to an annular rotor. The original factory was a small building where the wooden parts were made, while for many years the ironwork was put out to contract. The rotors were made in sections which could be shipped to the customer in a compact bundle and were easily erected in the field. The finished product bolted up into a stiff solid wheel that proved to be very safe, efficient and durable. In 1873, the Eclipse Windmill Company was organised which changed into Fairbanks, Morse and Company in the 1890s and continued manufacturing windmills into the 1950s. The Eclipse solid wheel windmill became one of the most popular types. In 1870, the first mill was installed for railway supply, a 6.1 m (20 ft.) diameter size at Elkhorn, Wisconsin. The largest mill at this time was

Figure 103. A former water-pumping mill now converted to generate electricity near Uden, North Brabant. When running, wind pushes against the fixed vane on the left which is counteracted by springs which hold the larger vane in the direction away from which the wind is blowing so that the sails remain facing the wind until it becomes too strong. The springs then yield, allowing the rotor to turn out of the wind (March 1992).

7.62 m (25 ft.) in diameter and versions soon ranged between 2.54 and 9.14 m (8½ and 30 ft.). By 1873 about 2,000 Eclipse mills had been built and the business continued to grow. In 1876, at the American Centennial Exposition in Philadelphia, the Eclipse windmill was awarded a gold medal and two years later received first prize at the Paris Exposition. A new factory had to be provided in 1880 where there were specialised sections like the foundry, planing mill, tank shop, hoop shop, machine shop, etc. The business in wooden railway water tanks and structures developed as the supply of water for railways grew, Pumps and cylinders were added to the product lines.[29]

The principles of governing developed by the Wheelers for their Eclipse solid wheel windmills were copied by scores of other makers. But there was a simpler method, using only a single vane, by placing the axle of the rotor to one side of the centre line of the main pivot on which the mill turned. Increasing wind pressure on the rotor turned the whole mill round and pushed the vane into the wind too. The vane could pivot sideways under the control of a weight or spring until equilibrium was established between the wind forces against rotor and vane. To stop the mill, once again the spring tension could be released or the weight raised to allow the vane to swing parallel to the rotor and so hold the blades out of the wind. Who developed this system is not clear. The firm of Williams, Smith & Co., at Kalamazoo, Michigan, were making the Manvel mill possibly from 1867 and their successors continued production of improved models until the 1920s.[30] In the windmill designed and patented by Palmer C. Perkins in 1869, the wheel was offset and so inclined away from the wind in stronger gusts. Here the vane assembly pivoted upwards through the action of the governing method and became its own weight to pull the rotor back once more as the wind lessened.[31] The Sandwich Enterprise Company fitted three weights connected to each other by chains. Initially only one weight was raised but as the wind increased and the vane turned round further, the chain from the first weight raised the second and so finally the third, giving a gradually heavier pull on the governing system. All three weights were raised only in very high winds.[32]

But vanes could be dispensed with entirely if the rotor were placed downwind of the mechanism and pivot. The principle was shown applied to a little windmill for scaring birds, by W. Emerson in 1754. If the vane is removed from a mill, the rotor will naturally blow around to the lee side of the tower. The weight of the rotor might be counterbalanced by a weight on an arm, which was often made into a decorative feature. While the blades no longer needed any mechanism to hold them into the wind, they still had to be feathered in gusts or gales. The most common method was with sectional tilting designs. As early as 1867, W.D. Nicholls of Chicago was patenting windmill improvements and in the same year some of his designs were being built by the Challenge Company at Batavia, Illinois. Althouse Wheeler and Company of Waupun, Wisconsin, were producing vaneless sectional mills which were patented in 1874 and 1876. Vaneless sectional mills assumed a predominant position among sectional types in the 1870s and 1880s and continued to be built well into the twentieth century.[33]

Some windmills, particularly those for driving machinery, were installed on the tops of buildings and even town tenements had windpumps to supply their residents with water. However, most were built for farmers on the tops of tall lattice towers. The first towers were built from wood. Individual parts were cut and drilled by the makers and only assembly was necessary on site. Compared with the heavy baulks of timber from which post and smock mills were built, the transport of these light sections for the windpump towers was much easier and the structures were much cheaper than traditional mills. These structures also presented less of an obstruction to winds and this must have improved efficiency. Later, wood was replaced by iron or steel. In larger windmill factories, the sections were cut and the holes punched with templates so that they were interchangeable. A tower of a certain height could be built up from parts of certain standard sizes and fastened together with bolts.[34] They would be fitted together in the shop before being taken apart for shipment so that they could be quickly erected on site. Sometimes they were assembled vertically but some were put together lying flat and then hauled upright. Four foundation blocks around the well were the only other requirements. Such mills could provide homestead farmers with a comparatively cheap supply of running water and one of the virtues which such men sought in their wives was the ability to climb a windmill tower to keep the mechanism oiled and greased.

> Within a short time after its introduction, the windmill became the unmistakable and universal sign of human habitation throughout the Great Plains area.[35]

Primitive windmills could be constructed from wood and waste materials quite easily and were the salvation of many farmers in years of drought because enough water could be raised to irrigate a small patch of land for families' vegetables. Such a mill might be built for as little as $1.50. Also at first it was possible for people like the Wheelers to start a windmill manufacturing concern with a few simple woodworking tools and build most of the parts of the mills themselves. New production methods introduced during the Industrial Revolution, followed by the 'American method' of manufacture, enabled demand to be met. But competition was always fierce and firms survived through cheap mass production.

How many firms were involved in this industry is shown by a rough count by decade of those listed in T. Lindsay Baker's book, *A Field Guide to American Windmills* as having started to build windmills.

Decade	Number of new firms
1850–60	15
1860–70	22
1870–80	96
1880–90	402
1890–1900	321
1900–10	159
1910–20	60[36]

Webb quoted figures from *The United States Statistical Abstract of the Census of Manufacturers, 1919*.

Date	Number	Employees	Value of products ($ U.S.)
1879	69	596	1,011,000
1889	77	1110	2,475,000
1899	63	2045	4,354,000
1904	53	1929	4,795,000
1909	34	2337	6,677,000
1914	31	1955	5,497,000
1919	31	1932	9,933,000[37]

These figures seem at variance, possibly because often the windmill was but one of a series of different products which a firm might make, but at least they begin to show the extent and importance of windmill production and how it peaked towards the end of the century. The distribution of manufacturers in 1919 was given as follows: Illinois, 8; Indiana, 4; Kansas, 3; Wisconsin, 5; Michigan, 2; and Iowa, Ohio, North Dakota, Texas and New Jersey, one each. Of these, all but one were located along the eastern edge of the Prairie and Great Plains area.

Production figures are harder to ascertain. The career of Allen S. Baker may be typical of how many started. He formed a partnership with Levi Shaw to manufacture a rotary steam tractor but, on its failure in 1873, the stockholders decided they must look for some alternative product to keep the small shop busy. Under Baker's supervision, the firm began producing windmills and sold about sixty in the following year. It was called the 'Monitor' in 1875 in honour of the iron-clad warship. In 1879, the firm was valued at $20,000 which had risen to $60,000 in 1883 when it was shipping as many as seventy windmills weekly.[38] This firm ceased supplying windmills in the 1960s. Some production figures have been preserved for the Aeromotor Company, which quickly became the largest of the American windmill manufacturers by introducing a new design of steel rotor based on scientific testing. The rate of growth in its early years was astonishing through low prices achieved by incredible production efficiency which beat its competitors. Only forty-five mills were sold in 1888, the first year; 2,288 were sold in 1889, 6,268 in 1890 and 20,049 in 1891; 60,000 were projected for 1892. By the turn of the century, the company had captured about half of the trade in the United States and it claimed to have over 800,000 windmills in service by the middle of this century.[39]

One of the few descriptions of windmill production was published in 1890 in *Farm Implement News* about the Aeromotor Company.

> It is believed that there is more manufacturing and business done per square foot in the ground that this building stands on than anywhere else in the . . . City. The work commences in the basement, where the heavy punching and cutting is done, and is completed as the material moves upward through the six floors of the factory. When it reaches the upper floor, it is ready to ship. There is virtually a stream of steel pouring into the basement and from thence on up through the building and out. If anything should stop the shipping a week the factory would have to shut down, because it could not store the output.
>
> One department is well adjusted to another, so that each can do its work in turn and allow the material to move on. There is constant shipping of Aeromotors which one week before were in pig metal, bar and sheet steel. The company did not think it could do business in this way, but the demands made upon it in the factory have made it a necessity, and whatever is a necessity in order to get goods out promptly is done, whether others consider it can be done or not.[40]

The credit for making the first all-steel windmill is ascribed to the patent of J.S. Risdon, Genoa, Illinois, which was granted in 1872 although the 'Iron Turbine' mill did not reach the market until 1876. The blades were shaped into 'buckets' which the patentee claimed gave 'more power than any other wind wheel of the same diameter on the market' and that, because it was made of steel, 'there is no wood . . . to swell, shrink, rattle and be torn to pieces by the wind'.[41] The manufacturers were Mast, Foos, and Company, Springfield, Ohio. The Iron Turbine had a profound impact on the history of American windmill design and manufacture because it showed that an all-metal windmill could do the work of a wooden one and that a metal mill could withstand severe winds as well as being repaired without great difficulty. However, these particular mills never achieved the popularity of their wooden counterparts. Other manufacturers also tried to introduce steel blades, but controversy over the merits of wood versus steel continued until the 1890s when steel had become relatively inexpensive to allow its use in increasing numbers of competitively priced windmills. While the numbers of wooden mills continued to increase until the First World War, the output of steel ones soon outstripped them by a long way.

This was partly attributable to the work of Thomas O. Perry, an employee of the U.S. Wind Engine and Pump Company. He spent several months in 1882 and 1883 carrying out meticulous tests on different designs of wind rotors about 1.52 m. (5 ft.) in diameter, which were the most exhaustive experiments since those of John Smeaton in Britain over 130 years earlier and bore similarities to his. Perry too constructed a rig with a horizontal rotating arm 4.27 m (14 ft.) long on the end of which the rotor could be secured (Figure 104). It was driven by an 80 h.p. Reynolds Corliss steam engine which could be governed to specially determined speeds. Perry developed instruments for recording measurements such as the revolutions of the wind rotor and a Prony brake was fitted on the windshaft. He carried out over five thousand tests on over fifty different types of rotors. Most of them had wooden vanes set at angles between 35 and 45 degrees. He soon found out that the majority of these violated Smeaton's principles because the total sail surface exceeded the total area of the annular zone and that a total sail area of 75 per cent was nearer the maximum to give the best results. He continued with experiments on steel blades and found that wide ones were better than narrow ones and that it was better to have fewer of them. One wheel with only six sails gave 2.5 times the efficiency of another with sixty.[42] After over a year's work, Perry had developed an entirely new steel rotor which was 87 per cent more efficient than the earlier wooden ones. Instead of having the thin, flat wooden blades nailed to wooden rims, his design consisted of concave sheet steel blades set at a specific angle to the wind fastened to steel rims and arms which gave enough strength but the least wind resistance.

After the U.S. Wind Engine and Pump Company had been employing him for fourteen months, the directors refused to accept his recommendations, possibly through fear of upsetting their customers with a completely new type of mill. While with this company, Perry met La Verne W. Noyes, an inventor and entrepreneur born at Genoa, New York. These two men joined forces in 1888 with other capitalists to manufacture the first scientifically designed windmill, which Noyes christened the Aeromotor, and hence the Aeromotor Company to manufacture it. Early American windmills had a single rotating shaft through which the rotary motion of the rotor was turned into a reciprocating one normally by a crank to give a direct drive to the pump. The shaft might turn in wood bearings made from maple soaked in oil, which often lasted longer than the iron shaft and it was claimed would never wear out. The connecting rod or pitman might be made from more oil-soaked wood and either this, or the reciprocating link to the pump rods, might be made deliberately weak so that this part would break and prevent damage to other moving parts should anything malfunction elsewhere. Bearings of all sorts were tried from mixtures of brass, gun metal or white metal as well as cast iron. Graphited 'oil-less' bearings were tried by some manufacturers to reduce maintenance but these did not run as

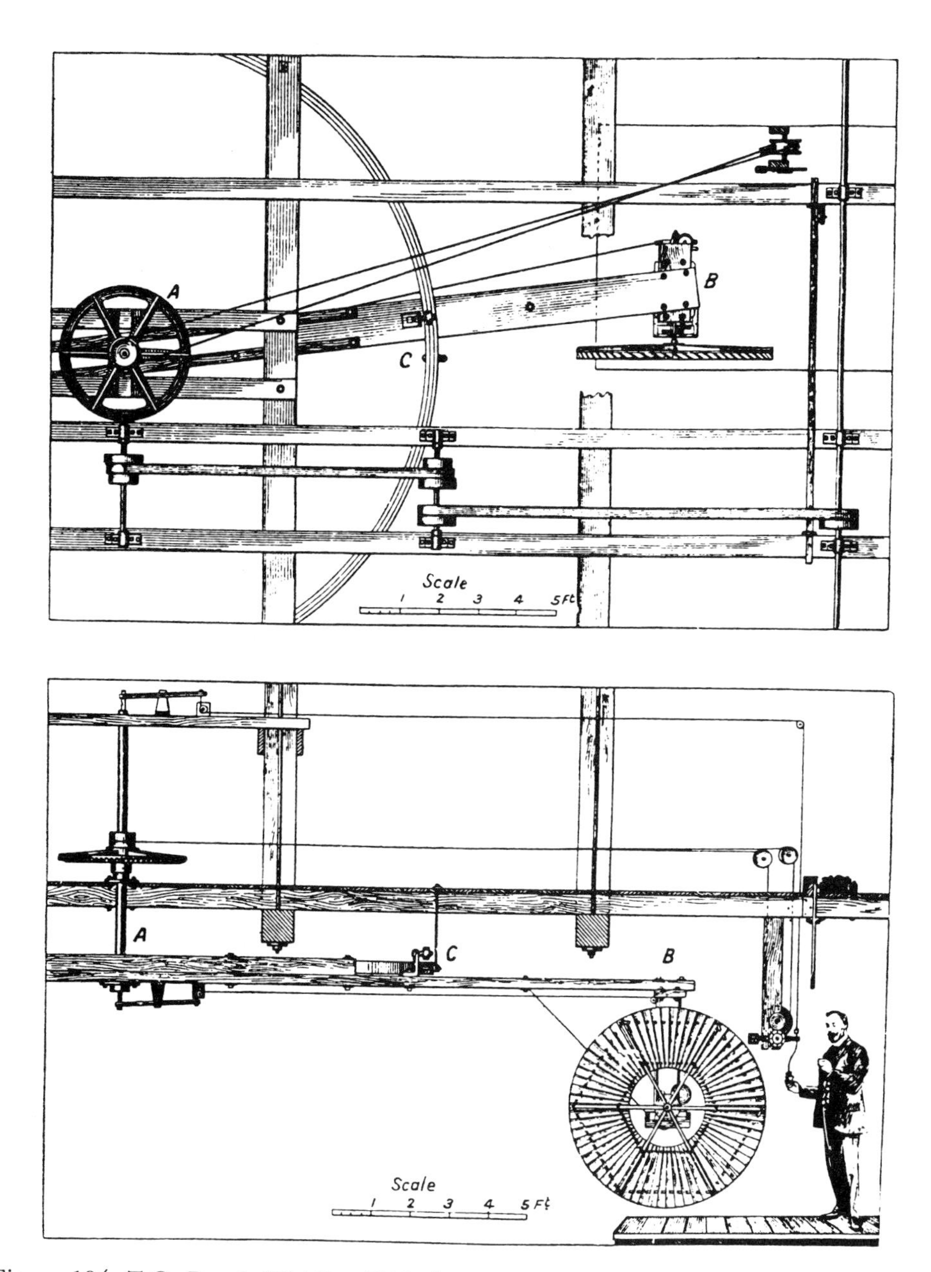

Figure 104. T.O. Perry's Whirling Table for testing windmills, plan and elevation (1882).

freely as some others. Ball or roller bearings in later designs after the turn of the century gave mills which rotated much more easily.

The rotor on Perry's Aeromotor windmill was made with concave sheet steel blades mounted on steel arms and rims, a type still known today among windmill makers as the Perry wheel. Its manufacture lent itself to mass production with special machinery and, as the design of windmills became more and more specialised with improved gearing and other details, so the American system of engineering was adopted. It was this sort of organisation which helped Perry to produce the Aeromotor economically. His wheel had the potential for revolving much more quickly but this meant it went too fast for a direct drive to the pump. Perry had to introduce gearing, which was advantageous in many ways. The wind forces on the windshaft were contained on one set of bearings while the pumping forces could now be taken on the bearings for the layshaft which operated the reciprocating pump rods. Reduction gearing between windshaft and lay shaft (Figure 105) meant that the wheel exerted much greater leverage on the pump rods so that the mill started in lighter winds. This, together with the improved form of blades, meant that the Aeromotor started pumping with winds of about 1.78 m/sec. (4 m.p.h.) which was a great advantage. Perry found

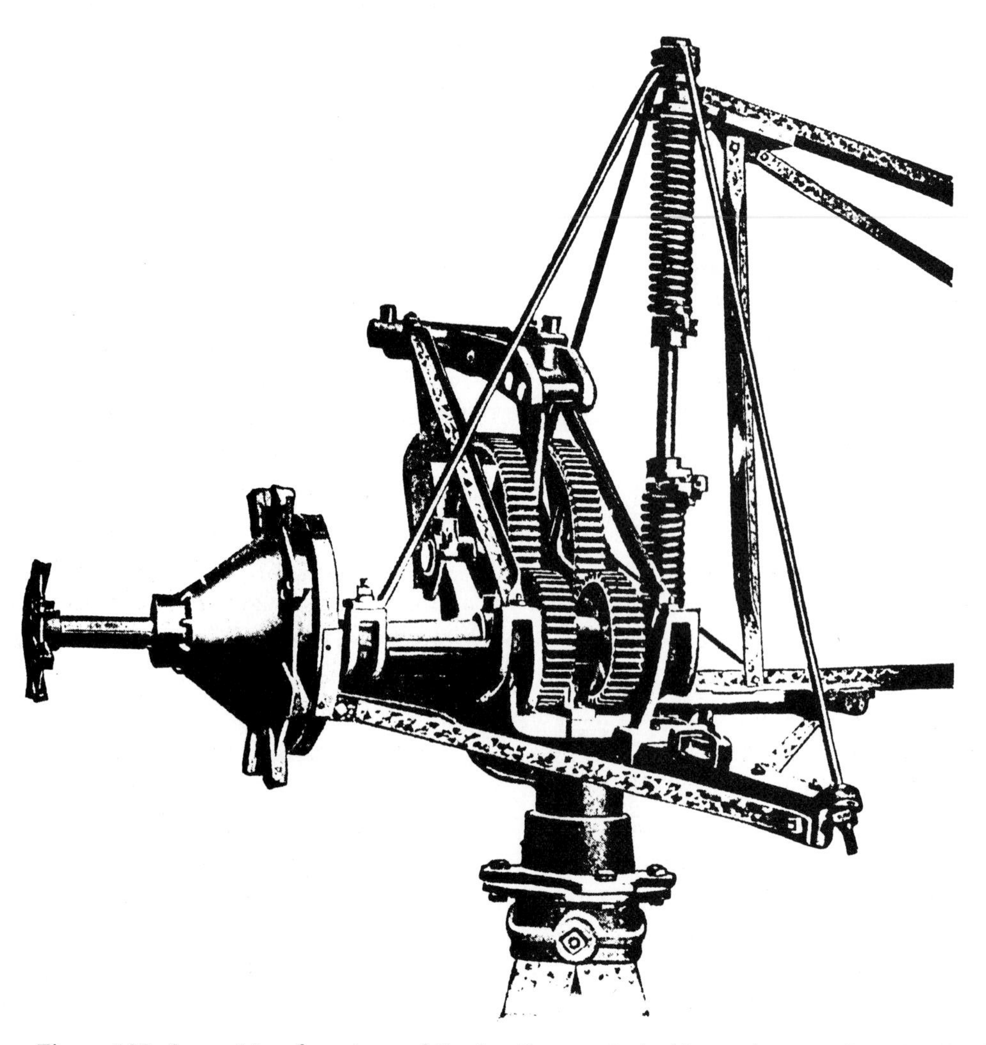

Figure 105. Stover Manufacturing and Engine Company's double gearing to reduce the speed between annular sails and the pump.

that the right proportion of gearing gave about three revolutions of the wheel to one stroke of the pump. He was able to design a pump with a long stroke which avoided the short, quick, jerky movement of direct stroke ones in high winds.[43] Aeromotor mills continued to be built into the 1980s.

The extra bearings for the two shafts and the gearwheels necessitated improved lubricaton. At first, the mechanism was exposed and oiling or greasing was done by people climbing the tower and filling the cups regularly. Sometimes the cups were made from glass, enabling the oil level to be checked from the ground. The task of climbing the towers in summer must have been unpleasant enough for anyone with a fear of heights, to say nothing of the dangers in the middle of a severe winter. The Aeromotor Company, and some others, devised tilting towers which had hinged sections so the top portion could be lowered to the ground to permit lubrication of the windmill parts, but these were not very successful. Other manufacturers introduced reservoirs to hold a large supply of oil which would need refilling infrequently. A valve in the bottom controlled the flow which was operated by a wire running down to the bottom of the mill. All the farmer had to do was to pull the wire to release more oil, which would last for several days. Probably the first of these oilers was introduced by the Burke–Bollmeyer Oiler Company in 1901.[44] However, there was often no way of knowing when the reservoir was empty and such oilers merely reduced the number of trips needed up the tower.

The important improvement was introduced in 1906 on the Little Giant windmill made by The Elgin Wind Power and Pump Company, which had been an innovator in design since its foundation in the 1880s. On their new windmill, bearings of both main and crankshafts were enclosed in oil reservoirs which were tightly closed after filling and were said to contain enough lubricant for twelve to eighteen months. The gearing and connecting rods were still left exposed. The success of this mill quickly led to further improvements with the first ever proper self-oiling windmill, in which all working parts were enclosed and protected from the elements (Figure 106). They operated in an oil-bath so that oil was continually circulated to all bearings. The crank gears lifted oil up to the main shaft and pinion bearings from where it was also taken up to the crosshead slides and bearings. The crosshead and slides were protected by a sheet metal hood which kept out dirt and weather and also prevented the oil from splashing everywhere. The new mill was called 'The Wonder' and someone commented 'Its a wonder you don't have to climb up and grease it every week'.[45] The Elgin Wind Power and Pump Company found the market very receptive to their new mill and sold many thousands of the original and later patterns until the Second World War. Self-oiling mills had a great impact on design both in America and elsewhere, for other makers found they had to follow suit quickly, even in England. In America itself by 1904, the windmill had become indispensible in many regions.

> America is the land of the windmill. Nowhere are so many mills to be seen, and nowhere is it seen used to advantage in so many ways. The windmill is something that the American farmer, never content with obsolete methods, wanted, and he got it. Now he would not know what to do without it; it is an integral part of American agricultural life.[46]

The American windmill spreads across the world

Continual improvement of windpumps over the years and the variable nature of their power source, together with the unique situation of each mill, made comparisons very difficult but some figures which show their capacities and capabilities are available. Writing in 1908, Robert Ball gave the following figures of performance and prices for windmills with annular sails erected on 12.19 m (40 ft.) steel towers in a 4.77 m/sec. (10 m.p.h.) wind.

Figure 106. A modern windpump at the Centre for Alternative Technology, Machynlleth, with off-set rotor and casing for automatic oiling to all moving parts (March 1992).

Diameter of wheel	Horsepower in 10 m.p.h. wind	Price (£ sterling)
12 feet	0.17	40
16	0.30	70
18	0.38	110
20	0.47	170
30	1.05	350[47]

The power produced by different wind speeds from a rotor of 4.88 m (16 ft.) diameter ranged from 0.02 h.p. at 1.78 m/sec. (4 m.p.h.) to 18 h.p. at 17 m/sec. (40 m.p.h.).[48] This shows how small is the power output. So when suitable internal combustion engines were placed on the market, farmers turned to them. A Petter two-stroke petrol/paraffin engine now in the possession of the author was sold in 1928 to a Cambridgeshire farmer for pumping water. It produces 1½ b.h.p. in a size of about 762 cm (2 ft. 6 in.) cube and can just be manhandled. Although the costs of fuel had to be considered, it could be run whenever necessary. For those people connected to the electric grid, a fractional-horse electric motor was a simpler, even smaller option.

The fullest information is contained in tests carried out in 1903 at Park Royal, London, by the Royal Agricultural Society.[49] These trials created a great deal of interest and the information was published in many places because this was the first time that a number of windmills had been brought together and tested at the same time under the same conditions. It was expected that these tests, carried out scientifically, would put the understanding of the windmill on a new footing. They presented an opportunity for evaluating different designs of rotor, governing, gearing and pumps but in the event they failed to come up to expectations. One reason may have been that each manufacturer supplied a complete pumping unit and the individual parts, such as the rotors, were not tested separately. This meant that it was impossible to discover whether the rotor was the best design while the pump might have been poor, or vice versa, and so the results were inconclusive.

The mills were run for two months, in March and April 1903, and had to pump against a head of 60.96 m (200 ft.) without any attention or repairs to be carried out during the trial period. Water was recirculated by the pump from a sump through a pressure valve which gave the effective head up to a storage tank from where the water could flow back to the sump. To everybody's surprise, it was a Canadian firm, Goold Shapley and Muir, Ontario, with mill number 3, which won first prize. In part, this was ascribed to the driving mechanism of their pump, which gave a long steady stroke through its mangle-motion transmission gear. This rotor had 18 blades and the others, except the last, 24. Thomas & Son, Worcester, won second prize with mill no. 7, and J.W. Titt, Warminster, came third with mill no. 8 and fifth with mill no. 16. Sixth was mill no. 17 of H. Sykes, London, with the highest number of blades at 48, based on a theory that the work was done on the leading edge. All rotors were 4.88 m (16 ft.) diameter.

	No. 3	No. 7	No. 8	No. 14	No. 16	No. 17
Speed ratio wheel to pump	5–1	2½–1	1–1	2–1	2½–1	1–1
Gallons per rev. of wheel	0.4	0.145	0.183	0.174	0.184	0.284
Revs of wheel per minute	20	30	22.18	31.2	18.5	16.9
Strokes per minute	4.01	12.01	22.18	15.60	7.41	19.91
Price of mill in £ sterling	70	77	61.7.6	79	74.10.5	106[50]

These figures show the speed of rotation of typical mills and also the rate of the waterpump.

The firm of J.W. Titt priced their Woodcock wind pumping engines in 1905 at the following rates.

10 ft. diam. Wind Wheel,	on wooden tower, 17 ft. high,	£27.10.0
	on steel tower, 25 ft. high,	£50. 0.0
12 ft. diam. Wind Wheel,	on wooden tower as above,	£38.10.0
	on steel tower as above,	£60. 0.0

Prices and power of their Simplex direct pumping mills and mills with gears for driving machinery ranged from £88 for a 1 h.p., 4.27 m (14 ft.) diameter to £440 for a 8 h.p., 12.19 m (40 ft.) diameter.[51] These prices may be compared with the salary of the Superintendent of the Waterbeach Level who, in 1903, received £80 per annum with free house and coal. The stoker of the Stretham engine, who also had a free house and coal, was earning fourteen shillings a week and was allowed one month's absence at harvest.[52]

A large market developed for these mills all over the world and in 1904 the *Chicago Tribune* wrote, 'The American windmill flies in every breeze the world over'.[53] Manufacturers in America were exporting some in the 1860s and 1870s. The trade grew to a considerable level in the 1880s and reached its peak in the years before the First World War. In 1889, the Williams Manufacturing Company reported that it had mills pumping water in South Africa, India, Australia, New Zealand, Mexico, the West Indies and South America. Even in Holland, annular sails on tall lattice towers began to make their appearance and challenge traditional mills (Figure 107). Within the next few years, it was reported that whole train-loads of windmills were being exported from the United States.[54] Robert Ball said,

Figure 107. American influence was felt even in the Netherlands, This is a large twin-vaned mill for pumping water in the Zaan area (August 1990).

> So widely used are these mills in great agricultural countries like Australia and the western parts of America and the Argentine that the manufacturers have found difficulty in keeping pace with the demand, notwithstanding the fact that there are several firms that can turn them out by the thousand and who can always find a ready market for them. They are cheap and are well made, will work without much attention, and are admirably adapted to pumping water and to any service not requiring a continuous output of power. They are easily erected, require very light foundations, and repairs and renewals are a small item, as they are now capable of withstanding the heaviest gales.[55]

Even if windmills had been developed as an economical source of power for pumping, other factors still had to be right, such as the wind and a supply of subterranean water at reasonable depth and in reasonable quantities. Some investigations were carried out in India in 1880 but it was felt that the wind in most areas was too uncertain and too variable.[56] Further trials were made in 1902 with an Aeromotor 4.88 m (16 ft.) diameter windmill mounted on a 21.33 m (70 ft.) tower with a 20.32 (8 in.) pump lifting water 9.84 m (25 ft.) high at Madras. Gearing between the rotor and pump was 3⅓ to 1. Careful observations were made on several days in conjunction with an anemometer fixed to the windmill tower. This showed that when the wind velocity exceeded 1.34 m/sec. (3 m.p.h.) a certain amount of work was done, but that a steady breeze of about 3.23 m/sec. (7½ m.p.h.) was needed to keep the mill in continuous motion.[57] It was reckoned that this windpump would be capable of irrigating 10 acres when the water had to be raised 7.62 m (25 ft.) Taking the capital costs of the windmill and allowing 10 per cent depreciation and 6 per cent interest, the monthly charge worked out at 25 rupees. To raise the same amount of water would need two pairs of good cattle at a cost of 67 rupees a month, so the windmill proved to be more economical. However, a survey of records from weather stations throughout the subcontinent showed that, while at Bangalore the wind movements were so much stronger that a similar windmill would do 60 per cent more work, throughout the greater part of India including the whole of Northern India, the winds were too weak to be of any practical value. 'At the great majority of stations, the wind forces are so slight that we may unhesitatingly say that wind-mills are of no value'.[58]

One of the most successful places for windpumps has been the Lasithi plateau, at an altitude of 850 m (2,000 ft.) on the island of Crete. The wind regime is favourable at around 6 m/sec. (13½ m.p.h.) during the growing season from April to October. Around 1890, a local carpenter built himself a windmill with an Italian pump and a rotor with typical Mediterranean cloth sails. During the 1920s, agricultural improvements were introduced and cultivated areas expanded which were irrigated by windmills. In the 1950s, there were nearly 16,000 windmills of various types at work in this region but most of them have now been replaced by diesel engines, for even here the wind proved to be too unreliable.[59] Improvement of similar windmills continues today for use in developing countries.

The winning of the Royal Agricultural Society trials by a Canadian manufacturer drew attention to the well established industry in that country. Around the shores of the Mediterranean, American exporters found themselves in competition with both British and German makers. In France, Ernest Bollée had patented in 1868 a novel and possibly unique design of annular sail mill, the Eolienne Bollée, which may have owed its origins to water turbine practice. It had two sets of blades, curved in opposite directions. One set, the stator, was fixed and its blades directed the wind onto the rotor. Another unusual feature was the surrounding of both sets of blades by a casing shaped in the form of a funnel to compress the air into the stator and through to the rotor. In this patent the sails were winded by a vane to the rear; it is possible that this feature did not prove satisfactory because no examples of éoliennes have been found in this period. It may be that Bollée was too early with his design in face of prejudice in favour of traditional mills but there could have been a design problem too. The sails of a mill with an ordinary annular rotor cause considerable turbulence to the rear and this might have been more pronounced in the Bollée design which affected the vane. In another patent in 1885, the mill was winded by a fantail in front of the stator and production continued from that date until the 1920s. The mill might be supported on a single cast iron column surrounded by a spiral staircase stayed by wires. More conventional lattice towers were also supplied.[60]

Great Britain was an early market for American exporters, where they had to compete against indigenous manufacturers. This started in the 1870s and, in the following decade, American mills featured at many agricultural shows because a number of British agricultural firms entered into agreements to act as agents.[61] With a strong demand for water-pumping windmills, a vigorous industry soon emerged in Britain itself. Comment has been made earlier about the visit of Americans to Henry Chopping which resulted in him taking out a patent in 1868 in conjunction with Frederick Warner. The firm John Warner, Cripplegate, London, continued to make mills with annular sails for many years in both water-pumping types and also ones with fantails and rotating shaft drives for machinery.[62] J. Thomas & Son, Broad Street, Worcester, built some 30,000 'Climax' wind engines up to their failure a little before 1972.[63] Some other firms were Duke and Ockenden, Littlehampton, Sussex; H.J. Goodwin, Ltd., Quennington, Gloucestershire; Hole and Robert, Warminster, Wiltshire; John Mullins, Bath, Somerset; E.H. Roberts, Deanshanger near Stony Stratford, Northamptonshire; and Wakes and Lamb, Ltd., Newark, Nottinghamshire.[64] This regional spread reflects the areas where these pumps were installed.

John Wallis Titt, Woodcock Ironworks, Warminster, Wiltshire, had been building windpumps since before 1887 and two of their mills, one geared 9.14 m (30 ft.) diameter and the other direct drive 4.88 m (16 ft.) diameter, featured in the Royal Agricultural Society trials in 1903. Uses for their mills in their catalogue a couple of years later were pumping water for water supplies, irrigation and drainage as well as driving machinery and electric lighting. They advertised a conventional design, the Woodcock, on either wooden or steel towers, the later being the more expensive. There was also the direct pumping 'Simplex' which was available in a geared form for driving machinery (Figure 108). In one version, this mill turned a modern bucket and chain pump made from iron for land drainage at Limavady, Co. Derry, Ireland, while another turned a scoopwheel. Driving electric dynamos was represented by a mill supplied to Simeon Colbeck, Esq., Boyle Hall, West Ardsley, Yorkshire. This mill was advertised with the remark,

Figure 108. This annular sailed mill driving reciprocating pumps is typical of the influence of American practice on later British windmill design. The shutters can be feathered while the mill is winded with a fantail because it is driving a rotating shaft to the pumps (J.W. Titt's Catalogue, 1905).

> This plant provides emancipation from the darkness and inconvenience hitherto felt by the dwellers in the outlying homes of the land, and supports Lord Kelvin's prophecy of the return to wind power, intelligently controlled, as one of the most economical powers for man's use, and, in addition to the convenience and beautiful effect of electric lighting, it is computed, from a health standpoint, that the general use of electric light would decrease the annual death rate by 40,000 in the United Kingdom, who now die from the effects of disease consequent on existing in an atmosphere where the oxygen is burnt up by gas, lamps, and candles.[65]

This vision of cheap electricity supplied by windmills was never fulfilled and the reasons why will be examined in the next chapter.

CHAPTER 10

Electric power from the wind

Stanley Jevons pointed out in 1865 that supplies of coal in Britain were finite and said 'With coal almost any feat is possible or easy; without it we are thrown back into the laborious poverty of early times . . . England's manufacturing and commercial greatness, at least, is at stake in this question'.[1] At Coalbrookdale, where so much of the early Industrial Revolution had started, coal seams were becoming worked out. Jevons thought there could be a future for electricity which was 'to the present age what the perpetual motion was to an age not far removed'.[2] Then an important use for electricity was discovered. In England, Joseph W. Swann demonstrated his lamp to the Newcastle upon Tyne Chemical Society on 18 December 1878 and, in America, Thomas A. Edison, on 21 October 1879, reached the end of a long search when his lamp, made with a piece of carbonised cotton sewing thread, glowed brilliantly for over forty hours. The two men joined forces in 1883. These lamps were suitable for lighting rooms in people's homes and were preferable to candles and oil or gas lamps while arc lights were too hot and bright. Today's vast electrical supply industry owes its origins directly to this demand in the 1880s and 1890s for domestic lighting.

The credit for being the first person to propose the use of the wind to generate electricity has been given to William Thomson, later Lord Kelvin, the eminent physicist and electrician, who in 1881 addressed the British Association for the Advancement of Science 'On the Sources of Energy in Nature Available to Man for the Production of Mechanical Effect'.[3] He saw difficulties in using windmills because their design and development had not advanced enough at that time and the capital costs were too great. In America, Alfred Wolff advocated their use in 1885,[4] and Joseph J. Freely, of Walpole, and George E. McQuestion, of Marblehead Neck, both in Massachusetts, set up small-scale, windpowered electricity generating plants during the later 1880s. These and other early trials concerned D.C. generation.

D.C. generation

One of the most prominent pioneers to take advantage of the new light was Charles F. Brush, Cleveland, Ohio. Around 1890, he erected a windmill to provide the power source for experiments as well as lighting his imposing mansion (Figure 109). The rotor was an annular sail, 17.10 m (56 ft.) in diameter, with fixed thin wooden blades. It was winded by a moveable rear vane 18.29 m (60 ft.) long with a fixed one at the side. All this was mounted on a rectangular tower rather like a post mill for the whole body turned to face the wind. Within this body was mounted the generating equipment which charged 408 secondary battery cells in the basement of his residence. The estate was equipped with 350 incandescent light bulbs, about 100 of which were in everyday use, as well as arc lights and electric motors.[5]

Another important pioneer was Professor Poul La Cour in Denmark. In 1891 he was placed in charge of an experimental windmill test station which was set up by the Danish Government at Askor, between Kalding and Esbjerg, and remained there until his death in 1907. He was interested in providing electricity for use on Danish farms and so his objectives were economical construction to

Figure 109. C.F. Brush's huge experimental electricity-generating windmill at Cleveland, Ohio, in the 1890s.

yield optimised performance. He erected a second windmill in 1897 which had slatted sails 22.8 m, (75 ft.) in diameter, each 2.5 m, (8¼ ft.) wide, which could be remotely controlled. It was winded by two fantails and drove two 9 kW (12 h.p.) dynamos on which the speed of the dynamo could be limited by adjusting the tension of the driving belts through a device he called a 'kratostat'. While this mill could develop 90 h.p. in a very strong wind, the shutters were set to spill the wind at 35 h.p. Inside the base of this mill, La Cour installed a machine room with the generators, physical and chemical laboratories and store rooms. One dynamo charged a 60 cell battery with a current of 150 volts, 50 amperes, for lighting and driving electric motors and provided a back-up in calms. The second dynamo generated 30 volts, 250 amps, which was used with electrolytic apparatus for decomposing water into oxygen and hydrogen, which were compressed into storage cylinders.

La Cour was the first person to undertake systematic investigations on wind power for electric generation for, in the physical laboratory, he carried out tests on different shapes of windmill sails with wind from an air propellor. He did not change the basic design of traditional sails and outlined the pattern he conceived necessary for his ideal mill:

(a) It should have four sails whose surface, particularly at the tips, should present the least resistance to rotation.
(b) The sails should be parallel-sided and of width one-fifth to one-quarter of the length.
(c) The sail surface should begin at a distance of about one-quarter of the length from the axis of rotation and the sail should occupy the remaining three-quarters.
(d) The sail profile near the tip should not be straight but bent convex to the wind at one-sixth to one-quarter of the width from the front edge.
(e) The weather angle should be 10° at the tip and should increase regularly to be 15° two-thirds of the length from the axis and 20° at one-third.
(f) The tip speed should be 2.4 times the speed of the wind from which the maximum amount of energy is to be extracted; it will thus be generally 6 or perhaps only 5 m/sec. (20 to 16 ft.).[6]

This concept differed little from that proposed by Smeaton. Although the mill was kept facing the wind automatically and also had self-reefing sails, La Cour found that it did not run steadily enough for electric lighting or charging accumulators. Therefore the 'kratostat' allowed the driving belts to slip to keep the dynamos at a steady speed. Yet he did succeed in popularising windpower for electro-agricultural purposes because soon there were thirty mills in operation.

The earliest reference to an installation in England concerns one in London where a lighting plant was being successfully operated by an American windmill in 1892. J.W. Titt's catalogue of 1905 refers to another erected for Simeon

Colbeck, Boyle Hall, West Ardsley, Yorkshire in February 1899.[7] On a hexagonal steel tower was mounted one of this company's Simplex geared 9.14 m (30 ft.) diameter annular sail rotors winded by a fantail. In the building at the bottom, the drive was taken through a 2.44 m (8 ft.) diameter pulley to a specially wound dynamo. The electricity was stored in cells made by the D.P. Battery Co., London, which had capacity equal to supplying the 109 lights of between 8 to 16 candlepower for eight days in winter. There was no back-up system, the accumulators providing the power in periods of calm. The windmill was situated nearly 200 m (200 yards) from the house and needed little attention, the main bearings being oiled roughly every fortnight. The plant was in charge of the gardener who had had no previous experience of electrical work and the installation had given great satisfaction for the first sixteen months of operation.

Around 1900, George Cadbury erected an American style windmill on his estate for pumping water and driving a dynamo. A clutch enabled the pumps to be thrown out of gear when it was desired to work the dynamo. However, the power from the wind proved to be unsuitable for charging batteries owing to its variableness and the uncertain speed of the windmill which was soon replaced by a gas engine. The windmill, 10.67 m (35 ft.) in diameter, was retained for pumping water, for which purposes it was admirably suited.[8] In England the last traditional style mill was erected in 1929 at St Margaret's Bay, overlooking the Straits of Dover in Kent. Sir William Beardsell intended that the windmill should generate D.C. electricity for the house to which it was attached. He had the mill constructed by the Canterbury firm of millwrights, Holman Brothers, who erected a smock mill with fantail and patent shuttered sails. Inside the body of the mill was the dynamo, which continued to be used until the house was requisitioned at the beginning of the Second World War.[9] The body of the mill was brown with the cap, sweeps, fan, brick base and stage being painted white.[10] These were systems on a medium scale but before wind generation could become popular, rotors, generators and accumulators had to be improved.

In 1888, A.B. Wolff claimed

> that the non-employment of the windmill in this connection, is also not owing to the often alleged fact that the rate of revolution of the windmill, according to the varying force of wind, is too irregular to run a dynamo for the purpose of charging a storage battery, or that the wind cannot be depended upon for a sufficient length of time per day . . . but that the electrical accumulators are not yet a satisfactory and assured success.[11]

Accumulators need a steady charging rate and the voltage should increase towards the end of the charge.[12] At George Cadbury's installation, the irregular speed of mill and dynamo caused the accumulators to wear out rapidly and the plates in them to buckle. Every expedient was tried to make this plant a success, to no avail.[13] So Wolff was right in one way about the accumulators but the charging rate could be modified by an appropriate design of dynamo. In 1909 Lancaster Burne said

> By using a specially wound dynamo, the plant may be entirely automatic. Such a machine has been in use for nearly ten years at an installation in this country [was it Colbeck's?], supplying the equivalent of one hundred 60 watt lamps – the sole source of power being a 30 ft. disc mill. The dynamo is shunt wound; but has a few turns in opposition to the shunt winding, with the result that as the voltage rises with increase of speed, more current passes through the series coils; this weakens the field and so prevents overloading.[14]

A dynamo for a windmill needed to have a large field and a small armarture so that the field strength went on increasing proportionately with increased speed. Such dynamos had to be specially designed for their particular windmill rotors and so were expensive.[15] The electrical connection had to be made when the dynamo was generating some power but not enough to cause a sudden jolt. The converse applied when the dynamo was being disconnected, or the windmill might suddenly speed up.

Small-scale designs were produced which proved reasonably satisfactory. In the later 1890s, Charles J. Jager of Boston, Massachusetts advertised 'Windmills adapted for Power' for small-scale generation as well as water-pumping mills.[16] In 1895, Nansen fitted a windmill on his ship *Fram* and Captain Scott followed this example on his Antarctic voyage in 1901, when he took another for lighting his ship SS *Discovery* to save fuel. Although a windmill should have been ideal, in this case it was not properly adapted.[17] The great growth in interest came with the introduction of radios and small electric lighting plants for homes and farms after the First World War. This was due not only to the desire to have electric light but also to have a system of charging accumulators for working radio sets, which helped overcome the isolation of the countryside. However, the manufacturers, particularly in America of annular-sailed mills, soon found that they rotated too slowly to drive generators and switched to faster two- or three-bladed types.

A wind rotor producing high torque at low speed will generate less power than one rotating at high speed producing low torque.[18] Electric generators, which usually rotate faster than the rotors, need the second type and machines based on these designs remained in common use from the 1920s in rural areas until the isolated communities or farms they served were reached by power lines of the electricity companies in the 1930s to 1950s. Around 1930, an Oxford University research team erected nine windmills on a site near Harpenden, Hertfordshire. These mills varied in design and size from American type windmills to medium-speed aerofoil machines with rotors from 2.4 m (8 ft.) to 9 m (29 ft. 6 in.) in diameter. The electrical energy they produced was measured and the lowest annual output was 83.5 kWh/m^2 and the highest 199 kWh/m^2, which would be commendable today. The inference was that a well designed axial flow windmill at a height of 10 m (33 ft.) should yield approximately 100 kWh/m^2 annually on a good site in inland England and more on the coast.[19]

During the Second World War, D.C. generators driven by aeroplane type propellers might be found in windy British sites mounted on masts 10 to 13 m (30 to 40 ft.) high. With wind speeds from 7 to 15 m/sec. (15 to 32 m.p.h.) they would produce 1,000 to 3,000 watts.[20] Manufacture of such machines reached its lowest ebb in the middle years of this century; but since then, partly through the increasing price of fossil fuels, and so electricity produced by them, and more recently the 'green' movement, the demand for small-sized, wind-powered generators has grown to such an extent that the market is once again flooded with a vast array of new systems.[21] In 1947, about 10,000 sets producing 1 kW or less were being sold annually in America and it was estimated that, by 1967, this market would have amounted to 200,000 kW.[22]

But, writing in 1931, H.P. Vowles summed up the problem with such windmills.

> Even supposing that the efficiency of such a mill could be increased rather more than threefold, so that when driving an electric generator an output 10 kilowatts was obtained, some 16,000 of such windmills would be required to equal the output of the giant steam turbine installed at Hell Gate Power Station, New York City . . . Taking these facts into consideration, also the difficulties due to uncertainty and irregularity of working, among which must be mentioned the necessity of using electric accumulators or other means to compensate for such irregularity, it will be seen that there is little prospect of developing power on an extensive scale by any known type of windmill with sails moving in a vertical plane.[23]

A.C. generation

As early as 1885, Sebastian Z. de Ferranti had realised the potential for generating alternating current at a high voltage in large central power stations and distributing it through a grid system to transformers which would reduce the current to a voltage suitable for domestic use. In 1888 he started to build the Deptford power station, which has been claimed to be the beginning of our present system of electric generation and distribution. His example has been followed across the rest of the world. To generate A.C. current and feed this into the grid, a windmill had to match the frequency of the grid and then maintain that frequency or speed. The nature of the A.C. cycles meant that the switch-in speed was high and the constant frequency meant that governing systems had to be improved. Therefore a reappraisal of windmill design was necessary.

The work on aerodynamics by Frederick Lanchester around 1900 passed unnoticed by his contemporaries in England as far as windmill design was concerned,[24] and there seem to have been very few people interested in this subject in Great Britain until after the Second World War. The development of the aeroplane, and more particularly the aeroplane propeller, does not seem to have stimulated research into windpower in this country although it did elsewhere. Towards the close of the First World War and immediately following it, scientists in France, Germany and Russia became interested in developing a modern theory of windpower in conjunction with the advances made during the war with the air propeller. Joukowsky, Drzewiecki, Krassovsky, Sabinin and Yurieff in Russia, Bilau, Prandt and Betz in Germany and Constantin and Eiffel in France laid the foundations of modern windmill theory. Eiffel, with his tower, had a structure which was useful for taking measurements of wind speeds at different heights while Betz was the first to show that no windmill could extract more than $^{16}/_{27}$ (about 59.3 per cent) of the energy passing through the area swept.[25] In 1926, the Russians, Sabinin and Yurieff published a design for a set of four blades which were partially streamlined, based on tests in a wind tunnel. Bilau, too, in Berlin had come to the same conclusion, that blades smoothed on the rear surfaces would improve the performance of a windmill.

Improved blades and rotors would enable a windmill to generate electric power on a larger scale for a village or to link up with the electric grid (Figure 110). Golding, writing in 1955, pointed out that medium-sized plant of 100 kW or 200 kW might be economic for islands, not because such windmills were cheap to build but because alternative methods of power generation might be more expensive, particularly in running costs.[26] Then a windmill feeding into a grid system might save other 'fuel' which would be freed for use when there was no wind. Golding cited a 1000 kW windmill which could help to reduce consumption of coal or oil. Windmills linked into a grid where some electricity was generated by hydraulic turbines could be just as advantageous if there were adequate reservoir capacity. In this instance, the windmill would generate electricity whenever there was sufficient wind and so save water being drawn out of the reservoir. The water would be kept for the time when there might be a drought. In this way, the windmill would have the effect of notionally increasing the rainfall or catchment area for the hydraulic system.[27]

In one way, the power generated by the wind is free because the air is free. But, while the source of energy is certainly free, the energy itself is thinly distributed in the air because its density is low. Therefore large volumes must be tapped to yield any appreciable output of power. Accordingly, the projected area of turbine blades to be driven by the wind must be greater per kilowatt of output than in the case of hydraulic or steam turbines. Equally, the structure to support this great rotor must be designed to withstand high wind and ice, which also increases the basic cost. The density of the air cannot be increased by supercharging.[28] This means that close attention must be given to the design of windmills so that they are as efficient as possible to give the maximum results. The plant required to capture this power and to convert it into usable form may have to be designed specially for a particular site and so may be expensive. The capital charges for interest and depreciation may raise the cost of each unit produced to an uneconomic level but it is generally true that the cost per horsepower of power-producing plant will decrease with increasing size. In 1947, it was estimated that a 1 kW windpowered generating set would cost from $300 to $600 per kilowatt. Hydraulic power in Vermont cost about $150 and a large wind turbine about $200.[29] In Great Britain just before 1955, medium-sized plants cost about 2.5 times those of large. Large-scale generation was economic with an annual mean wind speed of 4.77 m/sec. (10 m.p.h.) if the fuel component of generation by alternative means was greater than 1.5 pence per kWh. This figure was reduced to 0.25 pence per kWh with a wind speed of 11.18 m/sec. (25 m.p.h.).[30]

Figure 110. Modern two and three bladed design of wind turbines at Zurich, Friesland (August 1990).

One of the first places where it was shown that a windmill could be connected to a grid system was in the Netherlands. The 'Benthiuzer Bovenmolen', in the municipality of Hazerswoude to the east of Leiden, was being rebuilt after a fire in 1928. Not only was it re-equipped with A.J. Dekker's modernised sails but the scoopwheel was replaced by a single axial flow turbine pump. The pump had two driving pulleys, one operated by the windmill and the other by an electric motor. In periods of calm, the belt from the windmill was removed and that for the electric motor fitted. Normally there was only one belt in place at a time. H.A. Hoekstra carried out tests during the spring and summer of 1929 and, on a windy day that July, he fitted both belts so that the sails drove not only the pump but also the motor. When the motor had reached a fairly high speed, Hoekstra switched on the electric current.

> At first the kWh counter turned slowly in the right direction, but when the sails were struck by a squall of wind, the counter came to rest, than [*sic*] turned backward. As he considered the risk of damaging the gear rather high, HOEKSTRA propmly [*sic*] stopped this experiment, but he had sufficiently proven the possibility of generating alternating current with a windmill.[31]

In older style mills, the generating equipment was situated at the base of the tower where it was driven from the windshaft through a vertical shaft and then probably through belts. All of this driving mechanism absorbed power. In the new style wind turbines of the inter-war years, the generating equipment was nearly always placed as close as possible to the wind rotor, generally in the head of the mill. Shafting and gearing was kept to a minimum. One design which did not follow this was the Kumme, in 1920. It consisted of a six-bladed rotor which drove a vertical generator at ground level. While this seemed to be an attractive idea because the heavy weight of the electrical apparatus could be placed low down, it was found to be uneconomical in practice. When adequate provision was made to absorb the forces which cause a windmill to attempt to turn itself around the vertical drive shaft, throwing it out of yaw, as well as the problems of keeping the long flexible drive shaft aligned, it was found that having the generator on the ground was more costly than placing it aloft.[32]

A Texan, Dew Oliver, in 1926 erected a large wind generator at the San Gorgonio Pass in southern California. On a circular steel track he mounted a generator placed inside a large steel tube which opened out into a funnel at the end facing the wind to help increase the flow. The project failed through financial irregularities.[33] The French erected a more conventional two-bladed design at Bourget in 1929 with a high speed rotor 20 m (65 ft. 8 in.) in diameter. Overlooking the Black Sea at Yalta, the Russians produced a bold design in 1933 after two years of wind measurements. The tower was 30.48 m (100 ft.) high and the upwind rotor had twin 30.48 m (100 ft.) blades. The generation apparatus, contained in a streamlined pod in the head, was a 100 kW, 220 volt induction motor which reached its rated output at a wind speed of 11 m/sec. (24.6 m.p.h.). The maximum aerodynamic efficiency was 24 per cent at 30 r.p.m. In the same year, Honnef of Berlin suggested a tower 304.8 m (1000 ft.) high to support five 76.2 m (250 ft.) diameter turbines and generators mounted in a framework which could be rotated to face the wind. Such a structure would have presented formidable problems in stress analysis as well as the control of five turbines.[34]

The most important wind turbine was the one erected by Palmer Cosslett Putnam on Grandpa's Knob in Vermont. When the unit was phased in on 19 October 1941 at 6.56 p.m., it was run until 8.35 p.m. carrying loads which varied from 0 to 700 kW (940 h.p.). The wind speed varied from 7.15 to 11.62 m/sec. (15 to 26 m.p.h.) and was rather gusty.[35] This was the first time that a windmill had been connected directly on line with an alternating current grid system of an electricity supply company through a synchronous generator. It was designed to produce 1250 kW, enough to light a small town. The total weight was 250 tons while the tower was 33.53 m (110 ft.) high. The rotor had two stainless steel blades which were placed downwind at a height 36.58 m (120 ft.) above the ground. Each blade was 21.34 m (70 ft.) long by 3.35 m (11 ft.) wide and weighed 8 tons. Regulation was by pitch control through a fly-ball governor which was responsive to a speed change of 0.01 of 1 per cent through a hydraulic operating mechanism. The blades could cone downwind to reduce bending stresses at the roots. Its successful start followed years of intensive study into the nature of the wind itself, how this was affected by location, what was the best form of rotor and blades and what was the best design of structure to support and house a large-scale generating mechanism. In February 1943, a 609.6 cm (24 in.) downwind main bearing failed, which could not be replaced for two years owing to the war and then, on 26 March 1945, after the turbine had been restarted for a month, one of the eight ton blades snapped off near the root. This terminated the experiments, but much of great interest had been learnt.

When planning started, Putnam found that 'The state of our knowledge concerning the habit of wind in mountainous country was found to be meagre and uncertain.'[36] Therefore wind measurements were taken at various points around the Green Mountains and a steel tower 56.39 (185 ft.) high was erected on Grandpa's Knob itself. The intention of the 'Christmas Tree' was to measure the vertical distribution of the wind above a bare summit as well as the effects of the contour of the mountain. The right contour and aerodynamic characteristics could increase wind speed by 20 per cent or more. It was found that the height at which the maximum speed occurs over a summit such as Grandpa's Knob is about 60.96 m (200 ft.) and that the difference in average speed between the lower edge of the turbine disc at 15.24 m (50 ft.) and the upper edge at 72.24 m (237 ft.) is not over 15 per cent. Therefore little advantage could be gained by building much higher towers. To site the turbine so the blades came closer to the ground would mean that they entered an area of lesser wind speed and greater turbulence caused by ground friction. The diameter of the turbine rotor was related to the height of the tower, for too small a turbine on a tall tower penalised itself through higher construction costs while one too low entered the region of frictionally retarded winds. The windshaft was inclined down towards the rotor at the rear because this corresponded with the angle of the flow of the winds over the top of the hill, which were descending again when they hit the rotor. Two blades were chosen because the extra amount of power obtained from three did not justify the extra investment. It was thought that the three-bladed wind turbine held no dynamic advantage over a well designed two-bladed unit which was more economical to construct.

Control was by varying the pitch through the fly-ball governor. A design of plain blades was chosen after careful testing in conjunction with the Budd Manufacturing Company. Economical construction methods had to be balanced against aerodynamic efficiency. The airfoil section was dictated by structural considerations because the longitudinal spars had to be contained within it. The best blade, No. 15, was tapered with a curve which gave an output 95 per cent of the ideal. The largest tapered blade which it seemed reasonable to build, No. 13, with a straight taper, gave an output 92 per cent of the ideal. The best rectangular blade, No. 1, gave an output 88 per cent of the ideal. The Budd Manufacturing Company costed blades Nos 1 and 13 and, for a batch of 20, tapered blades were twice as expensive as the rectangular ones. The output from a rectangular blade is relatively insensitive to the amount of twist from root to tip and so No. 1 blade proved to be the most economical. Putnam said later that, at the time of the design of this turbine for Grandpa's Knob, there was little experience with flaps or spoilers at the tips of the blades to control the speed and that, from the knowledge he had gained, he would seriously consider this in another turbine. From the aerodynamic output so computed had to be subtracted losses caused by hub-windage and the tower shadow to give the gross input to the main shaft. Further deductions arose in gear losses, coupling losses, generator losses and the power required to operate servo-mechanisms and other auxiliary equipment. Operating costs were higher than had been expected but the potential for large-scale electric generation by wind-power had been demonstrated.

The Second World War halted further work on wind turbines but afterwards there was renewed interest in Britain which resulted in machines being installed on Costa Head, Orkney, in 1950, St Albans in 1953 and the Isle of Man in 1960. The one on Orkney had three thin blades, 5.18 m (17 ft.) long which were mounted at a height 23.77 m (78 ft.) above the ground. It operated for several years, feeding into the local diesel generated network. While there were teething troubles with these prototypes, valuable lessons were learnt but interest waned through the availability of cheap oil and the expectation of abundant low-cost nuclear power. The nascent wind energy programme was brought to an abrupt halt in the 1960s.

Demand by consumers for electricity is continuous but variable while energy from the wind is intermittent and variable, so that supply and demand seldom match. When feeding into a grid, the wind turbine normally will be meeting only a small fraction of the demand and the balance will come from other sources. So, until high winds prevail, most of the energy available through the rotor can be turned into useful electric current and the other generators governed to balance demand. Even so, some way of balancing the input of the wind with the output must be found because nearly all large-scale A.C. wind generators are of the constant speed type which must be phased in with the cycles of the rest of the grid. They may be either an induction generator with a squirrel cage rotor or a synchronous generator similar to a synchronous motor. Once a synchronous generator is phased in to its line, it will tend to stay in synchronism with the alternating current. To force it out of phase requires the application of a considerable torque, perhaps two or three times the rated torque.[37]

One way of altering the power being generated is altering the pitch of the blades to suit the wind speed and the load. Because the wind speed may change very rapidly, the pitch control mechanism must have an equally rapid response and must be actuated by the changes in wind speed and in the output.[38] With such blades, full advantage of the constant speed generator can, in theory, be realised over a wide range of wind velocities but it is not possible to design a generator that will cope with the highest velocities. Even with fixed blades and constant speed, power can be developed over a considerable range of wind velocities but with reduced efficiency except for a limited range.[39] Ideally for maximum efficiency, the wind rotor should operate as closely as possible to the best speed matching the velocity of the wind. In other words, the rotor speed should vary with the wind. There are two types of variable speed generators available, the wound-rotor induction generator and a synchronous generator with a frequency converter. However, these types have not found much favour with windmill designers although some researchers have inquired into the possible advantages of operating in a variable speed mode.[40] In the Netherlands, there has been erected recently an 80 kW wind turbine with hydrostatic transmission developed by Bohemen Energy Systems. Not only does this allow a variable speed rotor to drive a synchronous generator directly connected to the grid but the hydraulic power can be transferred to ground level where the heavy generating equipment can be situated, thus reducing the cost of the tower. The hydraulic pump is a low-speed high torque type with a constant stroke and a maximum of 52 r.p.m.[41]

There has to be sufficient wind to start any wind turbine. This may be a problem on those designed with narrow blades to give a high speed of rotation.

Sometimes the rotor may have to be started by an electric motor but generally the blades can be angled to an optimum position to give low-speed torque. Then there will be a wind speed at which the generator is designed to cut in and start generating current. Because light winds blow for the greater part of the time, the very best possible performance is needed here to increase the yearly output. Equally with high winds, there will be a cut-out speed when the turbine is shut down to prevent damage. Between these two there will be a wind speed at which the rated power of the generator is reached. This will be determined from the estimated demand or load and the wind regime. Above this value, it may be necessary to get rid of any extra power because this cannot be absorbed without threatening the safety of the machine.[42] This can be done by spilling the wind from the sails.

Since power is obtained largely by the outer third of the blade, this is also the most effective place to destroy it. Both very small control surfaces that break up the smooth flow of air over the blade near the tip or devices that greatly increase the drag will have a profound effect on power output. A small change in the angle of the tip of the blade will create a stall situation and reduce power considerably.[43] Therefore some designers have incorporated spoilers or sections at the ends of the blades which can be rotated to vary the pitch of that part alone (Figure 111). The central sections of the blades remain fixed and have the correct angle to give a good starting torque while the speed control is covered by the tip.

The oil crisis of 1973 renewed interest in wind energy because the price of oil rose dramatically. Now that this has fallen again, interest is waning once more because generation by diesel engine is cheaper; but modern technology is increasing the competitiveness of wind turbines through the computer and modern light-weight materials. The blades today may be built of laminated wood or fibreglass which can be moulded easily to the correct contours. Both are strong and light, and weather well. The profile can be designed to reduce noise from the wind and in the same way the gears too can be designed to run without whining. But the most important change has come through the introduction of computer control. Changing conditions of the wind, in both speed and direction, can be monitored continuously and matched to generation needs. Modern switchgear with, for example, thyristor equipment, has transformed control of the electrical apparatus.

In Britain, some use is being made of windpower to generate electricity. In 1982 on Fair Isle, a three-bladed 50 kW generator was installed with a rotor diameter of 14 m (46 ft.). Not only has this reduced electricity bills, as it has partially supplanted a diesel generator, but the islanders no longer fear the prospect of winter storms cutting them off from essential oil supplies. Similarly in Shetland, in 1988 a 750 kW wind turbine was phased in. It was built by James Howden & Co. Ltd. and, with a rotor diameter of 45 m (147 ft. 8 in.);

it is the largest three-bladed turbine in the United Kingdom. The control system, which is operated remotely from the diesel generating station at Lerwick 30 km (18.6 miles) away, works through spoilers at the tips of the blades and is based on two programmable logic controllers which regulate the operating and monitoring functions of the machine including the run up and shut down. It will generate with wind speeds between 4.91 and 25 m/sec. (11 to 56 m.p.h.), or up to a force ten storm. The load builds up until a wind speed of 12.96 m/sec. (29 m.p.h.) is reached, at which speed the full 750 kW is produced. This is maintained at higher wind speeds by the computer control system.[44] After the oil price rise in 1973, the cost of electricity in Shetland was two or three times greater than on the mainland. The original estimate for the whole of the wind turbine together with its control systems was £1,417,000. Taking into consideration maintenance and depreciation, the cost has worked out at 2.98 pence per kWh. This is now greater than the fuel for a diesel engine because, since the contract was placed for the wind turbine, the price of crude oil has fallen from $30 per barrel to less than $8 in 1986.[45] On Orkney, Burgar Hill

Figure 111. Edmund Lee would be surprized to know that his fantail was still being used over 200 years later. On this mill at Nieuwebildtdijk, Friesland, only part of the blades turn for speed control (August 1990).

has been chosen as the principal site for the development of wind turbine generators because it is one of the windiest places in Great Britain. Two machines were installed in 1983, one of 250 kW capacity and the other 300 kW. The smaller one was a prototype for a massive 3 MW machine which was completed in 1987, the most powerful wind turbine in Britain to date. It is a twin-bladed design 60 m (197 ft.) in diameter with twisting tips. The blades alone weigh 60 tons and the whole structure 900 tons excluding foundations. It can provide power for 2000 homes.[46]

Yet Britain has been lagging behind others in the development of wind turbines and most being installed today are of foreign manufacture. Owing to lack of interest among British producers, the Wood Green Animal Shelter, King's Bush Farm, near Huntingdon, turned to the Danish Company, Vestas, to supply a 225 kW machine. The Danes, with their lack of other power resources and low-lying countryside, have continued to investigate power from the wind since the time of Pour La Cour. The Vestas Company has built more than 3,500 modern wind turbines. The one at King's Bush Farm has a rotor 27 m (88 ft. 6 in.) in diameter with three variable pitch blades. Start-up wind speed is 3.5 m/sec. (4.5 m.p.h.), and cut-out 25 m/sec. (56 m.p.h.), with nominal output achieved at 13.5 m/sec. (30 m.p.h.). This turbine is fitted with a double-wound generator to give some output at low wind speeds. When using the small generator, the rotor speed is 32 r.p.m. and the generator produces 50 kW at 760 r.p.m. At higher wind speeds, a change is made to a rotor speed of 43 r.p.m. and a generator speed of 1,008 r.p.m. giving 225 kW. It has been estimated that the capital invested in the machine will be paid back in about three years through the electricity sold to the grid system.[47]

In Britain when the electricity supply industry was denationalised, provision was made for the installation of generation equipment using natural sources of power and this seems to have had some effect. A couple of examples can be mentioned of wind farms developed recently. In the centre of England, the main road from Ilkley to Skipton passes Chelker reservoir beside which are four wind turbines. But this site is eclipsed by the much larger one on the moors between Todmorden and Burnley. Presumably the wind is channelled along the pass between these towns, for on a small plateau high up on one side of the hill stand twenty-four wind turbines supplied by Vestas. Coal Clough wind farm is situated so that it is not visually intrusive and the noise level is minimal even when the turbines are all working. Other installations are in the course of erection.

In California, advantage has been taken of the mountainous divide between the coast and the desert to set up installations such as those in the Altamont Pass. In this state, some 17,000 wind turbines with a total capacity of over 1,800 MW have been installed.[48] The Dutch have revived their interest in windpower and wind turbines are springing up all over the country. In the southwest there are some situated on the enormous banks of the Delta project to keep out the sea, but to see large wind farms it is necessary to go the the northeast of the country. Near the small town of Franeker there is a group of ten 250 kW turbines, which opened in 1990 (Figure 112). A little further away is the more famous Sexperium wind farm, with eighteen larger turbines placed in three rows of six. All of these have three-bladed variable pitch rotors standing on tubular towers. Cows graze contentedly beneath them for even downwind there is little noise from the rotors or electrical apparatus.

Is there a future for power from the wind? Putnam thought that there were few sites in the whole of the world with a potential capacity of some 50,000 kW. He wrote,

> To provoke discussion, I will guess that somewhere on earth there are some forty-nine more sites like Lincoln Ridge, close to heavy load centres. Most of these sites would be found in such windy industrialized regions such as Scotland, northern Ireland, Iceland, Newfoundland, the Maritime Provinces of Canada, New England, other parts of the United States, southern Chile, New Zealand, Tasmania, and possibly high in the Italian Appennines, and in Scandinavia . . . In aggregate these fifty sites would amount to a potential market for large wind-turbines of 2,500,000 kilowatts . . . e.g. one hundred units of 2,500 kilowatts.[49]

Such wind turbines would be economic through their large-scale generation and low costs. The competiveness of smaller windmills in sites with less wind will depend upon the price of fuel, either oil or coal. Windmills will always need high capital investment because the density of the medium they employ is so low. Better generating and control equipment will be designed and production costs could be reduced if numbers of a standard design could be manufactured.

Figure 112. Cows graze beneath the 10 wind turbines in the group at Franeken, Friesland (August 1990).

In these ways, windpower could become more attractive. But should the future of power from wind be seen in these terms? Here is one form of power which does not contribute to the 'greenhouse effect' and it is on these grounds that future installation of wind turbines will most probably be justified.

Notes

CHAPTER 1 The wind

1 *The Holy Bible, Authorised Version Second Book of Kings*, Chapter 2, verse 11; *Book of Jonah*, Ch. 1, v. 4.
2 ibid., *Book of Ecclesiastes*, Ch. 1, v. 6.
3 ibid., *The Gospel According to St John*, Ch. 3, v. 8.
4 Golding, 1976, p. 502; Koeppl, 1982, p. 77.
5 Ball, 1908, p. 230.
6 Kempe, 1929, p. 1307.
7 Golding, 1976, p. 24.
8 Ball, 1908, p. 255; R. Wailes & J. Russell, 'Windmills in Kent', *T.N.S.*, Vol. 29, 1953–5, p. 233.
9 Golding, 1976, p. 41.
10 'Prinsenmolen'-Committee, no date, p. 47.
11 *Encyclopaedia Britannica*, 1910–11, article on 'Windmill'.
12 Golding, 1976, p. 54; Koeppl, 1982, pp. 17, 279.
13 van Dijk & Goedhart, 1990, p. 43.
14 Golding, 1976, pp. 24–5.
15 Koeppl, 1982, p. 99.
16 Stokhuyzen, 1962, p. 64.
17 Hills, 1967, pp. 29–31; Act of Parliament, 53 George III, L. & P., c. 81.
18 McDermott, 1978, p. 62.
19 Lancashire, Kenna & Fraenkel, 1987, p. 34.
20 de Little, 1972, p. 59.
21 'Prinsenmolen'-Committee, no date, p. 70.
22 Smeaton, 1796, p. 59.
23 'Prinsenmolen'-Committee, no date, p. 16.
24 Golding, 1976, p. 100.
25 Calvert, 1979, p. 80.
26 *Encyclopaedia Britannica*, 1910–11, article on 'Anemometer'.
27 Ball, 1908, p. 235.
28 Meteorological Office, 1981, p. 26.
29 Ball, 1908, p. 235.
30 Baker, 1985, p. 310.
31 Calvert, 1979, p. 81.
32 This table is derived from Ball, 1908, p. 218, Kempe, 1929, p. 1306–7 and Stokhuyzen, 1962, pp. 95–6.
33 Bean, 1987, p. 4.
34 Golding, 1976, pp. 22, 32; J.A. Griffiths, 'Windmills for Raising Water', *M.P.I.C.E.*, Vol. 119, 1895, p. 323.
35 Griffiths, 'Windmills', p. 324.
36 Ball, 1908, p. 255.
37 Lancashire, Kenna & Fraenkel, 1987, p. 34.

CHAPTER 2 The horizontal windmill

1 Landels, 1978, pp. 1226–7; Vowles & Vowles, 1931, p. 123.
2 *See* Baker, 1985, pp. 29–30 for the Jumbo mills and the comments about Hero's device by Dr J. Needham in R. Wailes, 'Some Windmill Fallacies', *T.N.S.*, Vol. 32, 1959–60, p. 108.
3 *See* Bathe, 1948; R. Wailes, 'Horizontal Windmills', *T.N.S.*, Vol. 40, 1967–8, pp. 125–45.
4 Needham, Vol. 4, 1954, p. 556; White, 1962, p. 86, note 7.
5 Vowles & Vowles, 1931, p. 124.
6 Klemm, 1959, p. 77. For a description of the wind in Seistan, see H.P. Vowles, 'An Inquiry into Origins of the Windmill', *T.N.S.*, Vol. 11, 1930–1, p. 8 written by Col. R.L. Kennion, British Consul there. 'The wind of Seistan! No one that has visited the country can ever think of it without

the consciousness that there the word "wind" acquired for him a new significance. For much as men receive new ideas about water after seeing an ocean storm, so a residence in these parts reveals new and unpleasant possibilities about air. There is . . . the wind of a hundred and twenty days, that howls through the land during the summer; the freezing but shorter lived winds which rage in the winter; all from the same point of the compass . . . Everywhere in Seistan are evidences of the wind – the wind and man's struggle against it. Hollows scooped out of the earth, clay buffs carved into fantastic shapes, dunes with horns pointing southwards, trees stunted and with a permanent list in the same direction, every building, the ruins even of buildings of the most ancient times, oriented southward.'

7 ibid., p. 78.
8 Wailes, 'Horizontal', p. 139. A detailed account of the surviving mills has been published by Harverson, 1991.
9 Harverson, 1991, p. 45.
10 Priveté Orton, 1953, Vol. 1, pp. 518–22.
11 Notebaart, 1972, p. 369.
12 Needham, 1954, p. 567.
13 White, 1962, p. 86.
14 H. Chatley, 'Far Eastern Engineering', *T.N.S.*, Vols 29 & 30, 1953–4 & 1954–5, *see* 'Discussion', p. 165.
15 ibid., p. 162.
16 Needham, 1954, p. 567.
17 Bennett & Elton, 1898, Vol. 2, p. 327.
18 *See* Golding, 1976.
19 White, 1962, p. 86.
20 Singer, Vol. 2, 1956, p. 626; *see* Besson, 1578.
21 Keller, 1964, p. 43.
22 Verantius, 1620.
23 *See* Blith, 1652.
24 Patent 1149, 14 March 1777, Stephen Hooper.
25 Rees, 1972, Vol. 5, p. 413; Gregory, 1807, Vol. 2.
26 Ireland, 1793, p. 8.
27 Wailes, 'Horizontal', pp. 135–6.
28 Wailes, 1954, p. 84.
29 Hopkins, 1970, p. 114.
30 Wailes, 'Horizontal', p. 136.
31 Baker, 1985, p. 14.
32 Keller, 1964, p. 45.
33 Mathé, 1980, p. 62.
34 Schofield, 1963, p. 74.
35 Patent 2679, 5 February 1803, Stephen Hooper.
36 Patent 2933, 3 May 1806, Stephen Hooper.
37 *See* McGuigan, 1978.
38 Nacfaire, 1988, pp. 12–15 & 112–15.
39 *The Independent*, 24 August 1990.

CHAPTER 3 The post mill

1 Gimpel, 1979, pp. 62–5.
2 Reynolds, 1983, p. 52.
3 Vowles & Vowles, 1931, p. 125; H.P. Vowles, 'An Inquiry into the Origins of the Windmill', *T.N.S.*, Vol 11, 1929, p. 9; contributions by M. Lewis and R. Gregory, 'The Origin of the Windmill in Western Europe', *The Society for the Protection of Ancient Buildings, Wind and Watermill Section*, Annual Windmill Meeting, 20 March 1993.
4 Forbes, 1958, p. 132; Kemp, 1983, pp. 55 & 63.
5 M.G. Huard, R. Wailes & Maj. H.A. Webster, 'Three Types of Windmills in Southern Britanny', *T.N.S.*, Vol. 27, 1949–51, p. 209.
6 ibid., p. 207.
7 Ramelli, 1976, Plate 132.
8 H.O. Clark & R. Wailes, 'Brake Wheels and Wallowers', *T.N.S.*, Vol. 26, 1947–9, p. 121.
9 Coulthard & Watts, 1978, pp. 18 & 39–42.
10 R. Wailes, 'Lincolnshire Windmills, Part I, Post Mills', *T.N.S.*, Vol. 27, 1949–51, p. 247.
11 The claim was made by M. Franklin, 'The Origins of Patents', *Chambers Journal*, January 1943, but Lynn White writes (1962, p. 87) 'The assertion of S. Lilley, *Men, Machines and History*, London, 1948, p. 211, that the European windmill first appears in a charter of 1105, is unwarranted: over a century ago Delisle showed that this charter must be a forgery.'
12 R. Wailes & Maj. H.A. Webster, 'Post Mills of the Nord', *T.N.S.*, Vol. 19, 1938–9, p. 130.
13 A.-M. Bautier, 'The Oldest References to Windmills in Europe', *T.I.M.S.*, 5th symposium 1982, pp. 111–19; Notebaart, 1972, p. 360.
14 See Kealey, 1987, for an enthusiastic account of early windmills in Britain.
15 Shaw, 1984, pp. 2–8.

16 Kealey, 1987, pp. 69 & 75.
17 Brunnarius, 1979, p. 1.
18 Kealey, 1987, p. 97; Gregory, 1985, p. 15.
19 Holt, 1988, p. 20.
20 ibid., p. 20.
21 ibid., p. 21.
22 Gimpel, 1979, p. 38.
23 Kealey, 1987, p. 8.
24 Bennett & Elton, 1898, Vol. 4, pp. 136 & 180.
25 Gimpel, 1979, pp. 67–8.
26 Clapham, 1950, pp. 77–8.
27 Flint, 1979, p. 1.
28 Holt, 1988, p. 34.
29 ibid., pp. 22–3.
30 ibid., p. 24.
31 Coulthard & Watt, 1978, p. 17. For an account of the construction of a windmill at Turweston, Buckinghamshire, in 1303, *see* J. Langdon, 'The Birth and Demise of a Medieval Windmill', *History of Technology*, Vol. 14, 1992, pp. 54–76.
32 Bennett & Elton, 1898, p. 240.
33 Watts, 1980, pp. 7–8.
34 Holt, 1988, p. 27.
35 Douch, 1963, p. 26; Guise & Lees, 1992, p. 6.
36 Holt, 1988, p. 116.
37 Gimpel, 1979, p. 37.
38 Caarten, 1990, p. 234; Notebaart, 1972, pp. 360–6.
39 Notebaart, 1972, pp. 360–4.
40 White, 1962, p. 87.
41 Bennett & Elton, 1898, Vol. 2, p. 231.
42 ibid., Vol. 2, p. 238; Dante, *The Inferno*, translation by J. Ciardi, Mentor Book, New American Library, U.S.A., 1954, Canto 34, lines 4–7.
43 Kealey, 1987, p. 39; Notebaart, 1972, p. 366.
44 Bennett & Elton, 1898, Vol. II, pp. 245 & 273.
45 Flint, 1979, p. 6; Pouw, 1981, p. 24.
46 For a description of Saxtead Green Mill, *see* Wailes, 1954, pp. 3–27 and Wailes, 1984.
47 R. Wailes, 'The Drive to the Stones in Windmills', *T.N.S.*, Vol. 31, 1957–9, p. 289.
48 H.O. Clark & R. Wailes, 'Notes on the Breton Sail and the Caviers or Hollow Post Mills of Anjou', *T.N.S.*, 1949–51, pp. 211–16.
49 A. Hirsjärvi & R. Wailes, 'Finnish Mills, Part III, Hollow Post Mills', *T.N.S.*, Vol. 44, 1971–2, pp. 99–106.
50 Gregory, 1985, p. 19.
51 Finch, 1976, p. 186.
52 R. Wailes, 'Suffolk Windmills; Part I, Post Mills', *T.N.S.*, Vol. 22, 1941–2, p. 41.

CHAPTER 4 Tower and smock mills

1 R. Wailes, 'Windmill Winding Gear', *T.N.S.*, Vol. 23, 1945–7, p. 28.
2 N.G. Calvert, 'Windpower in Eastern Crete', *T.N.S.*, Vol. 44, 1971–2, p. 137.
3 Notebaart, 1972, p. 375.
4 Wailes, 'Winding Gear', p. 28.
5 White, 1962, p. 161; Calvert, 1979, p. 9; Notebaart, 1972, p. 363.
6 Calvert, 'Eastern Crete', p. 137.
7 Reynolds, 1974, p. 97.
8 ibid, p. 95.
9 Douch, 1963, p. 47.
10 Notebaart, 1972, p. 382.
11 Pipe Rolls, 29 Edward I, *see* R. Wailes, 'Some Windmill Fallacies', *T.N.S.*, Vol. 32, 1959–60, p. 94.
12 Bennett & Elton, 1898, Vol. 2, p. 232; J.S.P. Buckland, 'Technical Notes on 16th and 17th Century London Windmills', *T.N.S.*, Vol. 60, 1988–9, p. 132; den Besten, Pouw & de Vries, 1991, Mill 24.
13 Singer, 1956, Vol. 2, p. 625.
14 Wailes, 'Fallacies', p. 94.
15 Flint, 1977, p. 31.
16 Ramelli, 1976, Plate 132.
17 Sipman, 1975, p. 526.
18 Caarten, 1990, pp. 219–24.
19 *See* Sipman, 1975, for a detailed account of this mill.
20 Reynolds, 1974, p. 78.
21 A. Titley & H.D. Haines, 'A Warwickshire Windmill and Some Notes on Early Millwrighting', *T.N.S.*, Vol. 28, 1951–3, p. 233f.; D. Ogden, 'Chesterton Windmill, Warwickshire, England', *T.I.M.S.*, 5th. Symposium, 1982; Wailes, 1954, p. 66f.
22 Coulthard & Watts, 1978, p. 25.
23 ibid, pp. 31 & 42.
24 Notebaart, 1972, p. 382.
25 Sipman, 1975, p. 526.
26 *See* den Besten, Pouw & de Vries, 1991, for some of these large mills.
27 McNeil, 1990, p. 250.
28 Hopkins, 1976, p. 18.
29 Gregory, 1985, p. 19.
30 Dolman, 1986, p. 3.

31 Scott, 1977, pp. 80–1.
32 Apling, 1984, pp. 61–2.
33 Flint, 1977, p. 31; M.H. Press, 'The High Mill, Great Yarmouth, Norfolk', *T.N.S.*, Vol. 26, 1947–9, pp. 275–6.
34 Wailes, 1954, p. 54f; Apted, 1986.
35 R. Wailes, 'Norfolk Windmills, Part 1, Corn Mills', *T.N.S.*, Vol. 26, 1947–9, p. 254.
36 Scott, 1977, p. 13.
37 Wailes, 'Norfolk, Part 1', p. 254.
38 R. Wailes, 'Suffolk Windmills, Part 2, Tower Mills', *T.N.S.*, Vol. 23, 1942–3, p. 51.
39 Mais, 1978, p. 197.
40 *See* Hopkins, 1976, p. 131 for West Somerton and Wailes, 1954, pp. 69 & 101 for Much Hadham mills.
41 Sipman, 1975, p. 527.
42 Hopkins, 1976, p. 76.
43 Apling, 1984, p. 7.
44 ibid, p. 8.
45 R. Wailes & J. Russell, 'Windmills in Kent', *T.N.S.*, Vols 29 & 30, 1953–5, p. 232.
46 Hopkins, 1976, p. 78.
47 Wailes, 1954, pp. 34–47.
48 Hopkins, 1976, p. 45.
49 Wailes, 'Norfolk, Part 1', p. 242.
50 Wailes, 'Suffolk, Part 2', p. 42.
51 Harte, 1849, Plate 24; A. Titley, 'Notes on Old Windmills', *TNS*, Vol. 3, 1922–3, p. 45; Ramelli, 1976, Plates 73 & 132.
52 'Prinsenmolen'-Committee, no date, pp. 17–20.
53 Sipman, 1975, p. 526.
54 R. Wailes, 'The Drive to the Stones in Windmills', *TNS*, Vol. 31, 1957–9, p. 290.
55 Flint, 1971, p. 10; Reynolds, 1974, p. 53; Ramelli, 1976, Plate 119.
56 Notebaart, 1972, p. 362.
57 Hunter, 1979, p. 44.
58 Notebaart, 1972, p. 360.
59 Hunter, 1979, p. 44.
60 Golding, 1976, p. 17.
61 ibid, p. 17; Stokhuyzen, 1962, p. 100.
62 Notebaart, 1972, p. 362.
63 See *Icelandic Journal of Henry Holland 1810*, ed. A. Warne. Hackluit Society, London, 1987, p. 217; Beenhakker, 1976.
64 G.D. Nash, 'An Introduction to the Windmills of Wales', *T.I.M.S.*, 6th Symposium, 1985, Paper 15; Guise & Lees, 1992, Foreword; Butt, 1967, p. 36.
65 Rhodes, 1962, p. 1.
66 Douch, 1963, pp. 54–5.
67 ibid., p. 13.
68 Apling, 1984, p. 7.
69 Wailes, 'Norfolk, Part 1', p. 232.
70 Flint, 1971, p. 3.
71 Brunnarius, 1979, p. 176.
72 MacLeod, 1988, p. 176.
73 Clapham, 1949, p. 186.
74 Quoted in Hills, 1970, p. 136.
75 J. Mosse, 'The Albion Mills, 1784–1791', *T.N.S.*, Vol. 40, 1967–8, p. 47f.
76 Hills, 1989, p. 64 f.
77 ibid., pp. 91–2.

CHAPTER 5 **The sails**

1 Diderot, 1763, Vol. 1, 'Agriculture', Plate 1 and 1959, Plate 16.
2 E.L. Burne, J. Russell, & R. Wailes, 'Windmill Sails', *T.N.S.*, Vol. 24, 1943–5, p. 147.
3 ibid., p. 147; Pouw, 1981, p. 24.
4 Burne, Russell & Wailes, 'Sails', p. 131.
5 ibid., p. 148.
6 R. Wailes, 'Windmills of Eastern Long Island', *T.N.S.*, Vol. 15, 1934–5, p. 124.
7 J.A. Griffiths, 'Windmills for Raising Water', *M.P.I.C.E.*, Vol. 119, 1895, p. 326.
8 Eggleston & Stoddard, 1987, p. 6.
9 Typescript, 'Introduction to Stevin's Essay on the Drainage-mills', p. 4.
10 Wailes, 'Eastern Long Island', p. 128; Pouw, 1981, p. 16.
11 *See* Linperch, 1727; van Natrus, Polly & van Vuuren, 1734; van Natrus, van Vuuren & Polly, 1736.
12 Stevin, 'Essay', pp. 7–9.
13 Bennett & Elton, 1898, Vol. 2, p. 560.
14 Stukeley, 1936, p. 39.
15 Belidor, 1782, Tome 3, pp. 30f.
16 Daumas, 1968, Tome 3, p. 5.
17 Emerson, 1794, p. 195.
18 Wolf, 1962, Vol. 2, p. 595.
19 Smeaton's papers to the Royal Society on wind and watermills were printed in 1796. For the experiments on windmills, *see* p. 38f.
20 ibid., p. 38.
21 ibid., p. 63.
22 Templeton, 1856, p. 111.
23 Smeaton, 1796, p. 48.
24 Wolf, 1962, Vol. 2, p. 596; Wailes, 1954, p. 92; Skempton, 1981, p. 44.
25 Ball, 1908, pp. 234–6.
26 Smeaton, 1796, p. 63.
27 Daumas, 1968, Tome 3, p. 7.

28 Rankine, 1861, p. 215.
29 'Inquiry into the Possibility of the Use of Wind Power for Irrigation in India', *M.P.I.C.E.*, Vol. 59, 1880, p. 409.
30 Griffiths, 'Windmills', p. 327.
31 Ferguson, 1803, p. 83.
32 Eggleston & Stoddard, 1987, p. 16.
33 Patent 471, 26 October 1724, John Brent.
34 Patent 561, 24 June 1738, John Kay.
35 Patent 615, 9 December 1745, Edmund Lee.
36 Buckland, 1987, p. 6.
37 R. Wailes, 'Windmill Winding Gear', *T.N.S.*, Vol. 25, 1945–7, p. 31.
38 Buckland, 1987, pp. 10 & 20.
39 Wailes, 'Winding Gear', p. 33.
40 Buckland, 1987, p. 19.
41 Skempton, 1981, p. 81.
42 Buckland, 1987, p. 19.
43 Desaguliers, 1743, Vol. 2, p. 537.
44 Patent 615, 9 December 1745, Edmund Lee.
45 Patent 1484, 11 June 1785, Robert Hilton.
46 Buchanan, 1841, p. 311.
47 Patent 1628, 15 November 1787, Thomas Mead.
48 Mayr, 1970, p. 102.
49 Dickinson & Jenkins, 1981, p. 220.
50 Patent 1706, 17 October 1789, Stephen Hooper.
51 Patent 609, 6 September 1744, William Perkins.
52 Patent 1588, 1 February 1787, Benjamin Heame.
53 Patent 615, 9 December 1745, Edmund Lee.
54 Buckland, 1987, p. 7.
55 Patent 1041, 21 April 1773, John Barber.
56 Patent 2438, 13 August 1800, Robert Sutton.
57 *See* Hills, 1989, p. 162f.
58 Patent 1399, 15 November 1783, Benjamin Wiseman.
59 Patent 1484, 11 June 1785, Robert Hilton.
60 Patent 1628, 15 November 1787, Thomas Mead.
61 Patent 2782, 14 September 1804, John Bywater.
62 H.E.S. Simmons, 'Note on Bywater's Windmill Sail Patent', *T.N.S.*, Vol. 36, 1963–4, p. 183.
63 Gray, 1806, p. 41.
64 Patent 1706, 29 October 1789, Stephen Hooper.
65 Gregory, 1985, p. 109.
66 R. Wailes, 'Notes on the Windmill Drawings in Smeaton's Designs', *T.N.S.*, Vol. 28, 1951–3, p. 239 & Plate 33.
67 Burne, Russell & Wailes, 'Sails', p. 152.
68 Patent 3041, 9 May 1807, William Cubitt.
69 Patent 2438, 13 August 1800, Robert Sutton.
70 Burne, Russell & Wailes, 'Sails', p. 153.
71 R. Wailes, 'Norfolk Windmills, Part 1, Cornmills', *T.N.S.*, Vol. 26, 1947–9, p. 247.
72 *See* Flint, 1979, p. 16; R. Wailes, 'Suffolk Windmills, Part 1, Post Mills', *T.N.S.*, Vol. 22, 1941–2, p. 49.
73 Flint, 1979, p. 16.
74 Varchmin & Radkau, 1988, p. 66.
75 McNeil, 1968, p. 76.
76 Skempton, 1981, pp. 65 & 254.
77 ibid., p. 255.
78 Calvert, 1979, p. 38.
79 E.L. Burne, 'Wind Power', *The Engineer*, Vol. 107, 19 March 1909, p. 286.
80 Patent 1041, 21 April 1773, John Barber.
81 Patent 6331, 8 November 1832, John Burlingham.
82 See R. Wailes, 'Suffolk Windmills, Part II, Tower Mills', *T.N.S.*, Vol. 23, 1942–3, pp. 45–6; Major, 1977, p. 1.
83 Patent 938, 19 March 1868, Frederick Warner and Henry Chopping.
84 Brunnarius, 1979, p. 172.
85 *Encyclopaedia Britannica*, 1910–11, article on 'Windmill'.

CHAPTER 6 Windmills for land drainage

The Netherlands

1 van Veen, 1962, pp. 18–19.
2 See Caarteen, 1990, and *Bijdragen en Mededelingen Betreffende de Geschiedenis der Netherlanden*, Netherlands, 1988, Vol. 103, Part 4.
3 Caarten, 1990, p. 212; Singer, 1957, Vol. 3, p. 302.
4 Blith, 1652, p. 56.
5 Bateson, 1710, p. 9.
6 J. Glynn, 'Draining Land by Steam Power', *T.S.A.*, Vol. 51, 1838, p. 10.
7 'Prinsenmolen'-Committee, no date, p. 20.
8 Patent 1023, 27 August 1772, Anthony George Eckhardt.
9 Harte, 1849, Plate 25; Young, 1799, p. 240.
10 *A.A.*, Vol. 26, 1796, p. 399.
11 'Prinsenmolen'-Committee, no date, p. 16.
12 *A.A.*, Vol. 26, 1796, p. 390.
13 van Heiningen, 1991, p. 13.

14 Singer, 1956, Vol. 2, p. 689.
15 *See* Blom, 1975, for a full description of this machine.
16 Typescript, 'Introduction to Stevin's Essay on the Drainage Mills', no date, p. 2; Singer, 1957, Vol. 3, p. 305.
17 Stokhuyzen, 1962, p. 120.
18 Singer, 1957, Vol. 3, p. 305.
19 ibid., 1956, Vol. 2, p. 689.
20 Kölker, 1990, p. 16.
21 Forbes, 1955, Vol. 2, p. 60.
22 R. Wailes, 'Upright Shafts in Windmills', *T.N.S.*, Vol. 30, 1955–7, p. 93.
23 W. Bosmon, 'The Origin of the Drainge Windmill', *T.I.M.S.*, 5th. Symposium 1982; Caarten, 1990, p. 45; 'Stevin', p. 2.
24 Caarten, 1990, pp. 54–6.
25 Singer, 1956, Vol. 2, p. 689.
26 Caarten, 1990, pp. 179–80.
27 Caarten, 1990, p. 168; Forbes, 1955, Vol. 2, p. 120.
28 Caarten, 1990, p. 140.
29 Stokhuyzen, 1962, pp. 28–30; Harte, 1849, Plate 30.
30 'Stevin', p. 2.
31 Notebaart, 1972, p. 382.
32 Singer, 1957, Vol. 3, p. 302.
33 Stokhuyzen, 1962, p. 18.
34 ibid, p. 18; leaflet, *Schermerland Museummolen*.
35 van Veen, 1962, p. 47.
36 'Prinsenmolen'-Committee, no date, pp. 179–82.
37 de Little, 1972, p. 58.
38 Eggleston & Stoddard, 1987, p. 23.
39 Stokhuyzen, 1962, pp. 96–7.
40 This is based on information received from Ir.L. Monhemius and Ir.W. Badon Ghijben. *see also* K. van der Pols, 'Early Steam Pumping Engines in the Netherlands' *T.N.S.*, Vol. 46, 1973, p. 13 and K. van der Pols, *De Ontwikkeling van het Wateropvoerwerktuig in Nederland, 1770–1870*, for a full account of the installation of the early pumping engines in the Netherlands.
41 Farey, 1827, p. 266.
42 Hills, 1967, p. 51.
43 R.L. Hills, 'The Cruquius Engine, Heemstede, Holland', *I.A.*, Vol. 3, No. 1, February 1966, p. 9.
44 H. W. Lintsen, 'From Windmill to Steam Engine Waterpumping; Innovation in the Netherlands during the 19th century, *Technik in Einzel Darstllungen* c. 1988. p. 334.
45 Hills, 'Cruquius', p. 10.

England

1 Williams, 1970, pp. 11–14.
2 ibid, pp. 95 & 112; Coulthard & Watts, 1978, p. 48.
3 Gregory, 1985, p. 71.
4 Skempton, 1981, pp. 139–49.
5 For general accounts of fen drainage, see Darby, *The Medieval Fenland* and *The Draining of the Fens*. A more recent, shorter account can be found in Summers, *The Great Level, A History of Drainage and Land Reclamation in the Fens*. Hills, 1967, covers the machinery and agricultural changes in the switch from wind to steampower.
6 H. Bradley, 'Discourse in the State of the Marshes or Flooded Lands', 1589, Lansdowne, 60/34.
7 *See* Hills, 1967, for a full discussion on agriculture in the Fens.
8 *See* Acts of Parliament 1 Anne c. 11, 27 February 1702, 11 & 12 William III, c. 22, 11 April 1700, 12 Anne, c. 7, 28 May 1714, 24 George II, c. 22 May 1751. Complete lists of the Acts may be found in Priestley, 1969.
9 Kinderley, 1751, p. 13.
10 Dodson, 1665, p. 17.
11 Burrell, 1642, p. 5.
12 Gooch, 1813, p. 246.
13 Hills, 1967, pp. 62 & 67.
14 Golborne, 1792, p. 33.
15 Bedford Level Corporation, Records of Earlier Courts of Sewers, 19 Richard II; *see also* 21 Henry VIII.
16 Harleian Ms., 701, pp. 25 & 30.
17 *See* Acts of Parliament, e.g. Haddenham Level, 13 George I, c. 18 (1727) and Waterbeach Level, 14 Geo. II, c. 24 (1741).
18 Harleian Ms., 5011/41 and B.L.C., Records of Earlier Courts of Sewers, 15 James I.
19 *See* Kirkus, 1959.
20 State Papers, Domestic, Elizabeth, 106/62.
21 ibid.
22 Lansdowne Ms., 110/3.
23 S.P., Dom. Eliz., 241/114.
24 ibid., 106/62.
25 Acts of the Privy Council, 26 June 1580, pp. 68–9.
26 Lansdowne Ms., 41/46, p. 181.
27 ibid., 41/49, p. 195.
28 ibid., 41/46, p. 181.
29 ibid., 41/49, p. 199.
30 ibid., 41/49, p. 203.
31 S.P., Dom. Eliz., 219/73.
32 Lansdowne Ms., 41/49, p. 203.
33 ibid., 46/53, p. 113.

34 Harleian Ms., 5011/41; *also* Ely Diocesan Records, Fen Drainage Bundle, p. 282.
35 Casaubon, 1850, Vol II, p. 866.
36 B.L.C., Records of Earlier Courts of Sewers, 15 Jam. I.
37 15 Charles II, c. 17, 1660.
38 B.L.C., London, 3 July 1701.
39 B.L.C., Ely, 23 July 1725.
40 B.L.C., London, 8 May 1663.
41 ibid., 18 August 1664.
42 ibid., 8 December 1680.
43 ibid., 1 December 1681.
44 ibid., 1 December 1681.
45 ibid., 19 and 31 August 1693; *see also* 28 May 1696 for an attempt to set up a mill near Whittlesey.
46 ibid., 31 August 1693.
47 ibid., 7 March 1698.
48 ibid., 7 March 1698.
49 B.L.C. as Commissioners of Sewers, 1700, 1701 & 1708.
50 ibid., 5 March 1700.
51 ibid., 28 March 1701.
52 Wells, 1830, Vol. I, p. 435.
53 ibid., p. 429.
54 ibid., p. 431.
55 B.L.C., London, 6 February 1706.
56 ibid., 11 July 1699.
57 ibid., 30 November 1699.
58 ibid., 11 July, 30 November, 7, 14 & 26 December 1669 and 27 February 1700.
59 ibid., 5 December 1701.
60 ibid., 5 December 1701.
61 ibid., 25 March 1703.
62 ibid., 25 March 1703.
63 B.L.C., Ely, 9 April 1703.
64 B.L.C., London, 26 December 1699.
65 ibid., 18 February 1702/3.
66 ibid., 14 December 1721.
67 B.L.C., Ely, 8 April 1727.
68 Badeslade, 1725, p. 93.
69 Miller, 1889, intro., p. iii.
70 B.L.C., London, 16 March 1726.
71 Badeslade, 1725, p. 94.
72 *J.H.C.*, 2 March 1726/7.
73 B.L.C., Ely, 9 April 1726.
74 ibid., 7 April 1727.
75 B.L.C., London, 2 March 1726/7.
76 13 Geo, I, c. 18, 1727.
77 *J.H.C.*, 20 February 1728; *see also* 7 February 1728.
78 ibid., 19 & 28 February 1728.
79 ibid., 20 February 1728.
80 B.L.C., Ely, April 1726, 1729 and 1731.
81 Armstrong, 1725; Labelye, 1745, which includes T. Badeslade, *A Scheme for Draining the Great Level of the Fens Called Bedford Level*, 1729.
82 *J.H.C.*, 6 March 1737.
83 11 Geo. II, c. 34, 1738.
84 14 Geo. II, c. 24, 1741.
85 Act for Haddenham Level, 13 Geo. I, c. 18, 1737.
86 ibid.
87 Wailes, 1954, p. 81.
88 Neale, 1748, p. 14; Defoe, 1724, Vol. II, p. 151.
89 L. Gibbs, 'Pumping Machinery in the Fenland and by the Trentside', *M.I.P.C.E.*, Vol. XCIV, 1888, p. 267.
90 ibid., p. 266.
91 Wheeler, 1897, p. 379.
92 Wailes, 1954, p. 80.
93 53 Geo. III, L. & P., c. 81, 1813.
94 Heathcote, 1876, p. 44.
95 Ely Diocesan Records, Haddenham Level Account Books, 1739–41 and 1743–5.
96 Young, 1799, p. 240.
97 J.A. Clarke, 'The Great Level of the Fens', *J.R.A.S.*, Vol. 8, 1848, p. 102.
98 Lansdowne Ms., 41/46 (1584).
99 Waterbeach Level Commissioners Minute Book, 1775 & 1793.
100 Littleport and Downham Commissioners, District Order Book, 23 May 1810.
101 Dodson, 1665, p. 22.
102 Wheeler, 1897, p. 379.
103 Bateson, 1710, p. 9.
104 Waterbeach Level Account Book, 1815–16.
105 Gooch, 1813, p. 67; Young, *A.A.*, Vol. 4, p. 278, Vol. 16, p. 464, & Vol. 36, p. 90.
106 J. Rennie, Letter Books, Vol. 11, p. 342, 6 December 1820.
107 Dempsey, 1854, p. 12.
108 Boulton & Watt Collection, R. Wild to J. Watt, 28 November 1789.
109 ibid., J. Watt to W. Swansborough, 6 August 1814.
110 Nickalls, 1793, p. 4.
111 J. Rennie, 'Report on the Improvement of the Outfall of the Vernatt's Drain in the County of Lincoln', 1818; *see The Reports of Civil Engineers on the Improvement of the Drainage of Deeping Fen and of the Outfall of the River Welland*, Shaw & Sons, London, 1852.
112 Figures supplied by the Meteorological Office.
113 B. & W. Col., J. Watt to W. Swansborough, 6 August 1814.

114 North Level District Letter Book, 1 February 1820.
115 Labelye, 1745, p. 55.
116 B. & W. Col., J. Watt to W. Swansborough, 18 August 1814.
117 Gooch, 1813, p. 239.
118 Stone, 1794, p. 23.
119 Waterbeach Level Records, Report of Flood, February 1900.
120 Dodson, 1665, p. 23.
121 Littleport and Downham District Minutes, 13 October 1812. Other double lift mills mentioned in these Minutes are Littleport, 31 October 1827 and Westmoor, 27 November 1828.
122 Waterbeach Level Order Book, 27 May 1814.
123 Gibbs, 'Pumping Machinery', p. 276.
124 Young, *A.A.*, Vol. 44, p. 278.
125 Young, 1799, p. 240.
126 Young, *A.A.*, Vol. 44, p. 278.
127 Wheeler, 1897, p. 323.
128 R. Wailes, 'Essex Windmills', *T.N.S.*, 1957–9, p. 162.
129 J. Rennie, Letter Books, Vol. 7, p. 412, 28 January 1813; Stone, 1794, p. 157.
130 Young, *A.A.*, Vol. 44, p. 281.
131 Gooch, 1813, p. 239.
132 Warburton, 1851, 21 June 1769.
133 ibid., 28 November 1770 & 18 April 1771.
134 Young, *A.A.*, Vol 43, p. 543.
135 J. Rennie, Letter Books, Vol. 11, p. 340, 6 December 1820.
136 Maxwell, 1793, p. 24.
137 J.A. Clarke, *Fen Sketches*, London, 1852, p. 246.
138 Neale, 1748, p. 14.
139 Clarke, 'Great Level', p. 94.
140 Walker & Craddock, 1849, p. 440.
141 Wheeler, 1897, p. 301.
142 Clarke, 'Great Level', p. 118.
143 Moxon, 1878, p. 12.
144 Gibbs, 'Pumping Machinery', p. 277.
145 Wailes, 1954, p. 80.
146 For a full discussion about the introduction of steam drainage to the Fens, *see* Hills, 1967.
147 *See* Lambert, 1960.
148 *See* note 112 for Fen figures; for Dutch, *see* Stokhuyzen, 1962, p. 97 and for the Broads, Caton, 1976.
149 R. Wailes, 'Norfolk Windmills: Part II, Drainage and Pumping Mills Including those of Suffolk', *T.N.S.*, Vol. 30, 1955–6, p. 159.
150 Flint, 1979, p. 90f, for a full description of the Herringfleet mill.
151 Wailes, 'Norfolk Windmills', p. 159.
152 ibid., pp. 174–5; Snelling, 1983.
153 Smith, 1990, pp. 56–9.
154 Wailes, 1954, p. 72.
155 Smith, 1990, p. 12.
156 Wailes, 1954, p. 80.
157 Wailes, 1982 for the Berney Arms Mill.
158 Miller & Skertchly, 1878, pp. 163–4.
159 Gibbs, 'Pumping Machinery', p. 270.
160 Heathcote, 1887, p. 12.
161 ibid., p. 17.
162 Wailes, 'Norfolk Windmills', p. 167.
163 ibid., p. 173.
164 Wailes, 1954, p. 72.

CHAPTER 7 Windpower for industry

Sawing

1 den Besten, Pouw & de Vries, 1991, p. 91; Reynolds, 1970, p. 175.
2 Kölker, 1990, pp. 16 & 77; Dobber & de Vries, *c.* 1990, pp. 12–13.
3 Stokhuyzen, 1962, p. 16.
4 ibid., p. 53.
5 Dobber & de Vries, *c.* 1990, pp. 15–17.
6 Hopkins, 1976, p. 149.
7 Patent 255, 1687, Duke of Albemarle.
8 Smiles, 1967, pp. 165–6.
9 Gregory, 1985, p. 25.
10 Shaw, 1984, p. 95.
11 ibid., p. 99.
12 ibid., p. 442.
13 ibid., p. 441.
14 Apling, 1984, p. 18–19.
15 Finch, 1976, pp. 161, 180 & 228; R. Wailes & J. Russell, 'Windmills in Kent', *T.N.S.*, Vol. 29, 1953–4, p. 225.
16 Brunnarius, 1979, p. 172; Flint, 1979, p. 50.
17 A. Hirsjärvi & R. Wailes, 'Finnish Mills, Part II, Mamsel or Smock Mills', *T.N.S.*, Vol. 43, 1970–1 and 'Finnish Mills, Part III, Hollow Post Mills', *T.N.S.*, Vol. 44, 1971–2.

Crushing and pulping

1 Kölker, 1990, p. 17.
2 Schofield, 1963, p. 74.
3 Stokhuyzen, 1962, pp. 79 & 83–5.
4 R. Wailes & Maj. H.A. Webster, 'Post Mills of The Nord', *T.N.S.*, Vol. 19, 1938–9, pp. 138–41.
5 'Introduction to Stevin's Essay on the Drainage-mills', typescript, p. 3.

6 Harte, 1849, Plates 26–30; Kölker, 1990, p. 17.
7 Apling, 1984, pp. 7 & 19.
8 *See* Hartlib, 1650.
9 Smith, 1990, p. 65; Snelling, 1983, pp. 14–15.
10 J.K. Major, 'An Inventory of Windmills in Northumberland and Durham', *I.A.* Vol. 4, No. 4, November 1967, p. 33.
11 Skempton, 1981, pp. 12 & 79; Wailes, 1954, Plate IX.
12 R. Wailes, 'Notes on the Windmill Drawings in Smeaton's Designs', *T.N.S.*, Vol. 27, 1949–51, p. 240.
13 Skempton, 1981, pp. 28 & 255.
14 Wailes, 'Smeaton's Drawings', p. 241.
15 Shaw, 1984, p. 210.
16 Dickinson & Jenkins, 1981, p. 164.
17 Copeland, 1972, p. 5.
18 Weatherill, 1971, pp. 23–4.
19 Patent 487, 5 November 1726, Thomas Benson.
20 Patent 536, 14 January 1732, Thomas Benson.
21 Skempton, 1981, p. 12.
22 Boucher, 1968, p. 31.
23 Smiles, 1874, pp. 147–8.
24 Gregory, 1985, pp. 24–5 & 108.
25 Wailes, 1990, p. 2.
26 Stokhuyzen, 1962, p. 78; van der Woude, 1983, p. 799.
27 Helmers, *c.* 1990, contains drawings of barley hulling equipment.
28 Shaw, 1984, p. 379.
29 I am indebted to Dr J. Kingma for this information.

Papermaking

1 Pels & Voorn, no date; van der Woude, 1983, pp. 489 & 799.
2 Valls, 1978, p. 2.
3 Hills, 1988, p. 21.
4 Hunter, 1978, p. 480.
5 Patent 307, 24 March 1692, Thomas Hutton.
6 Shorter, 1957, p. 49, note 114.
7 Hills, 1988, p. 56.
8 Hunter, 1978, p. 162.
9 Rees, 1972, Vol. 4, p. 78; Thomson, 1974, p. 159.
10 Bolam, 1965, pp. 2–3.
11 van Natrus, Polly & van Vuuren, 1969, Plates XV & XLX; Harte, 1849, Plate XLI.
12 Hills, 1988, p. 15.
13 van Natrus, Polly & van Vuuren, 1969, Plate XIX; Harte, 1849, Plate XL.
14 Hills, 1988, p. 62.
15 Pels & Voorn, no date.
16 R.L. Hills, 'De Schoolmeester: A Dutch Papermaking Windmill', *I.A.*, No. 3, August 1973, pp. 304–8.
17 *See* Pels & Voorn, no date; de Iongh, 1934, pp. 97–109.
18 Patent 220, 10 July 1682, Nathaniell Bladen.
19 Patent 242, 11 October 1684, Christopher Jackson.
20 Shorter, 1957, p. 52.
21 ibid., p. 209.
22 ibid., p. 52.
23 Hasted, 1790, Vol. 3, p. 391.
24 Gregory, 1985, p. 26.
25 Hunter, 1978, p. 551.
26 Hills, 1988, p. 52.
27 R.L. Hills, 'Some Notes on the Matthias Koops Papers', *The Quarterly*, August 1990, p. 5.
28 Hills, 1988, p. 162.

Mining

1 Agricola, 1950, p. 201.
2 ibid, pp. 205–6.
3 D'Acres, 1930, pp. 31–2.
4 Ramelli, 1976, Plate 73.
5 Galloway, 1969, p. 77.
6 Douch, 1963, p. 9.
7 Pryce, 1778, pp. 150–1.
8 Zonca, 1607, various pages.
9 D'Acres, 1930, p. 2.
10 ibid., p. 2.
11 ibid., p. 16.
12 Pryce, 1778, p. 152.
13 Griffin, 1977, p. 22.
14 Lewis, 1970, pp. 90–1.
15 Griffin, 1971, p. 26; Galloway, 1969, pp. 53–4.
16 Griffin, 1977, p. 104; Griffin, 1971, pp. 32 & 90.
17 MacLeod, 1988, pp. 54 & 81.
18 Blith, 1652, p. 40.
19 Lansdowne Manuscripts, 110/3.
20 Patent 174, 27 February 1674, John Johnson.
21 Patent 243, 12 November 1684, Nathan Heckford.
22 Patent 307, 24 March 1692, Thomas Hutton.
23 Galloway, 1969, p. 65.
24 ibid., p. 66.
25 Coulthard & Watts, 1978, p. 81.
26 Douch, 1963, p. 9.
27 ibid., p. 11.
28 Hills, 1989, pp. 16–20.
29 Rolt & Allen, 1977, p. 49.

30 ibid., p. 44.
31 Patent 471, 26 October 1724, John Brent.
32 C. Matschoss, 'A Holograph Letter of Newcomen', *TNS*, Vol. 2, 1921–2, p. 115.
33 Patent 561, 24 June 1738, John Kay.
34 Shaw, 1984, p. 63.
35 ibid., p. 64.
36 Duckham, 1970, p. 325.
37 Galloway, 1969, p. 78.
38 ibid., p. 78.
39 Shaw, 1984, p. 70.
40 Duckham, 1970, p. 77.
41 Rothes Manuscripts, 40/85/5.
42 ibid., 40/85/1.
43 ibid., 40/85/2.
44 Duckham, 1970, p. 80.
45 Griffin, 1977, pp. 46–7; Griffin, 1971, p. 97.
46 Duckham, 1970, p. 86.
47 Rothes Manuscripts, 40/85/2.
48 Duckham, 1970, p. 78.
49 ibid., p. 77.
50 Galloway, 1969, p. 78.
51 Shaw, 1984, p. 379.
52 Douch, 1963, p. 11.
53 Trevithick, 1872, Vol. I, p. 62.
54 ibid., Vol. II, p. 298.
55 Patent 547, 12 September 1732, Anthony Parsons.
56 Patent 643, 9 May 1794, Richard Langworthy.
57 Patent 1460, 15 January 1785, Christopher Gullett.
58 ibid.
59 Douch, 1963, p. 11.
60 Patent 1588, 1 February 1787, Benjamin Heame.
61 Douch, 1963, p. 12.
62 ibid., p. 10.
63 Morgan Rees, 1975, p. 145 and G.D. Nash, 'An Introduction to the Windmills of Wales', *T.I.M.S.*, 6th Symposium, 1985, Paper 15.
64 Douch, 1963, p. 10; Guise & Lees, 1992, pp. 135–6.
65 Hills, 1989, p. 70.
66 *Daedalus*, 1948; Letter from W. Wilkinson to James Watt, 11 October 1788; *Skanska Industriminnen*, Skanes Hembygdsförbund, Kristianstad, 1978, p. 66.
67 Agricola, 1950, pp. 284, 286 & 299.
68 Shaw, 1984, p. 79.
69 Patent 643, 9 May 1749, Richard Langworthy.
70 Buchanan & Cossons, 1969, p. 127.
71 Douch, 1963, p. 10.
72 Mott, 1983, p. 31; Patent 505, 21 November 1728, John Payne.

Textiles and agriculture

1 Patent 143, 3 March 1664, Abraham Hill.
2 Patent 288, 2 January 1692, Charles Moreton & Samuell Weale.
3 Shaw, 1984, p. 171.
4 ibid., p. 172.
5 ibid., p. 174.
6 van Natrus, van Vuren & Polly, 1736, Plate X; van der Woude, 1983, p. 799.
7 Apling, 1984, p. 19.
8 Smith, 1965, pp. 76–7.
9 Clapham, 1951, p. 154.
10 Focsa, 1971, p. 317.
11 van Natrus, Polly & van Vuuren, 1734, Plate XXIV.
12 Harte, 1849, Plates XLVI & XLVII.
13 *See* Smeaton, 1796.
14 *See* Hills, 1970, for the textile industry *and* Hills, 1989, for an outline history of the stationary steam engine.
15 Fussell, 1985, p. 156f.; Shaw, 1984, pp. 156 & 158.
16 Shaw, 1984, p. 161.
17 ibid., p. 499.
18 Bawden, Garrad, Qualtrough & Scatchard, 1972, p. 185.
19 *See* illustration in I. Sangster, *Sugar and Jamaica*, Nelson, London, 1973, p. 17.
20 Dickinson & Titley, 1934, p. 131.
21 Hughes, 1959, pp. 6 & 13.

CHAPTER 8 The demise of the traditional windmill

1 Crocker & Kane, 1990, pp. 72 & 91.
2 Fairbairn, 1861, p. 276.
3 Clarke, 1986, p. 16.
4 Coulthard & Watts, 1978, p. 20.
5 Voller, 1897, p. 190.
6 ibid., p. 4; Watts, 1983, p. 8.
7 Simon, 1953, p. vii.
8 Watts, 1983, pp. 14–15.
9 van der Woude, 1983, p. 798.
10 van Veen, 1962, p. 79.
11 ibid., p. 84.
12 van der Woude, 1983, p. 613.
13 Griffiths, 1979, p. 75.
14 ibid., p. 76.
15 ibid., p. 77.
16 J. Kingma, 'De Introduktie van de Stoommachine in de Zaanstreek', *I.A.N.*, No. 21, 1986, p. 60f.

17 H.W. Lintsen, 'From Windmill to Steam Engine Waterpumping; Innovation in the Netherlands during the 19th Century', *Technik in Einzel Darstellungen*, *c.* 1988, p. 330.
18 ibid., p. 340; *see* Archive Polder Nieuwkoop, 1756–1923, Nr. 847, 848.
19 ibid., p. 335.
20 Pouw, 1981, p. 49.
21 'Prinsenmolen'-Committee, no date, pp. 31–2; Pouw, 1981, p. 51–2.
22 'Prinsenmolen'-Committee, no date, p. 19.
23 ibid., p. 37.
24 ibid., p. 44.
25 Pouw, 1981, p. 53.
26 'Prinsenmolen'-Committee, no date, p. 127.
27 Stokhuyzen, 1962, p. 97.
28 'Prinsenmolen'-Committee, no date, p. 85.
29 Pouw, 1981, p. 55.
30 ibid., pp. 63 & 88.
31 ibid., pp. 100–5.
32 'Prinsenmolen'-Committee, no date, p. 184.
33 A description of the equipping of the Traanroeier mill can be found in *'De Traanroeier', Energie – Monument van het Verleden* and the articles by F.D. Pigeaud and Rex Wailes.

CHAPTER 9 The windmill for pumping water and water supply

Britain

1 Ramelli, 1976, Plate 73.
2 Butt, 1967, p. 37.
3 Wailes, 1954, p. 83.
4 Finch, 1976, pp. 212 & 261.
5 Dolman, 1986, p. 5; Gregory, 1985, pp. 74–5.
6 Shaw, 1984, p. 535.
7 Wilson, 1955, p. 39, 'Busino, 1617–19, Diaries and Dispatches of the Venetian Embassy'.
8 MacLeod, 1988, p. 81.
9 McNeil, 1968, p. 76.
10 Apling, 1984, p. 19.
11 Patent 547, 12 September 1734, Anthony Parsons.
12 Household, 1969, p. 85.
13 R. Wailes & J. Russell, 'Windmills of Kent', *T.N.S.*, Vol. 29, 1953–5, p. 234.
14 Finch, 1976, pp. 203, 242 & 184.
15 Clarke, 1916, p. 145.
16 Ball, 1908, p. 331.

America

1 Hunter, 1979, p. 47.
2 Baker, 1985, p. 123.
3 Torrey, 1976, p. 72.
4 ibid., p. 74; Beedell, 1975, p. 123.
5 Stott, *c.* 1975, pp. 14–15. *See also* Comp & Hoeft, 1976; R. Wailes, 'Windmills of Eastern Long Island', *T.N.S.*, Vol. 15, 1934–5.
6 Cochran & Miller, 1961, pp. 107–8.
7 Webb, 1976, *see* Preface.
8 ibid., p. 174.
9 Hounshell, 1984, p. 18.
10 Webb, 1976, p. 230.
11 ibid., p. 335.
12 ibid., p. 240; Baker, 1985, p. 55; *W.G.*, Vol. 1, No. 1, Winter 1982, p. 4, 'Many of the smaller ranches in Texas have from 100 to 350 mills continually pumping water'.
13 Webb, 1976, p. 22.
14 Baker, 1985, p. 4.
15 E. Lancaster Burne, 'Wind Power', *The Engineer*, Vol. 107, 19 & 26 March 1909, pp. 286 & 308.
16 Lancashire, Kenna & Fraenkel, 1987, p. 14.
17 Burne, 'Wind Power', p. 286.
18 Baker, 1985, p. 25.
19 ibid., p. 424.
20 ibid., pp. 5–7; Webb, 1976, p. 337f.
21 Baker, 1985, p. 8.
22 Burne, 'Wind Power', p. 307.
23 Baker, 1985, p. 26.
24 ibid., p. 26.
25 Burne, 'Wind Power', p. 307.
26 Baker, 1985, p. 26.
27 Patent 692, 28 March 1855, Joseph Peabody.
28 U.S. Patent 68674, 10 September 1867, L.H. Wheeler.
29 Baker, 1985, pp. 10–12; F.G. Hobart, 'History of the "Eclipse" windmill and Its Production', *W.G.*, Vol. 1, No. 1, Winter 1982, pp. 6–8.
30 Baker, 1985, p. 268.
31 ibid., p. 268.
32 ibid., p. 285.
33 For the bird scarer, *see* Emerson, 1794, and Baker, 1985, pp. 118 & 120.
34 Ball, 1908, p. 265.
35 Webb, 1976, p. 336.
36 Baker, 1985: these figures are reckoned from Appendix A.
37 Webb, 1976, p. 337.
38 Baker, 1985, p. 70.
39 ibid., p. 38.
40 ibid., p. 64.
41 ibid., p. 34.

42 ibid., pp. 36–7; Ball, 1908, pp. 305–15.
43 Baker, 1985, pp. 37–8.
44 ibid., p. 42.
45 ibid., pp. 43–4.
46 *W.G.*, Vol. 1, No. 1, Winter 1982, p. 3.
47 Ball, 1908, p. 332.
48 ibid., p. 319.
49 *J.R.A.S.*, Vol. 64, p. 174; *Encyclopaedia Britannica*, 1910–11, article on 'Windmill'; 'The Royal Agricultural Society's Windmill Trials', *The Engineer*, Vol. 95, 1 May 1903, p. 431.
50 Burne, 'Wind Power', p. 308.
51 Wallis Titt, 1905, pp. 7 & 9.
52 Hills, 1967, p. 138.
53 *W.G.*, Vol. 1, No. 1, Winter, 1982, p. 3.
54 Baker, 1985, p. 101.
55 Ball, 1908, p. 261.
56 'Inquiry into the Possibility of the Use of Wind Power for Irrigation in India', *M.P.I.C.E.*, Vol. 59, 1880, p. 409.
57 Chatterton, 1904, pp. 68–70.
58 ibid., pp. 81 & 83.
59 van Dijk & Goedhart, 1990, p. 14; Hoogervorst & van 't Land, 1983, gives a thorough account of the origins, development, economics and decline of these mills.
60 *See* Gaucheron & Major, 1985, which covers the history of this fascinating type of windmill. I wish to thank Kenneth Major for drawing my attention to it.
61 Baker, 1985, p. 103.
62 *Encyclopaedia Britannica*, 1910–11, article on 'Windmill'.
63 J.K. Major & H. Major, 'Wind Engines, A Necessary Study', *T.I.M.S.*, 3rd Symposium, 1973, p. 96.
64 Baker, 1985, p. 455; J. Sawtell, 'Annular Vaned Windpumps', Corfield, 1978, p. 29.
65 Wallis Titt, 1905, p. 18; for some notes on the life of J.W. Titt, *see* J.K. Major, 'Wind Engines', The Rolt Memorial Lecture, *I.A.R.*, Vol. 14, No. 1, Autumn 1991, p. 59.

CHAPTER 10 Electric power from the Wind

1 Jevons, 1865, pp. viii & ix.
2 ibid., p. 119.
3 A.B. Wolf[f], 'Letter', 15 January, *The Engineer*, Vol 65, 3 Feb, 1888.
4 *See* Wolff, 1885.
5 Baker, 1985, p. 45.
6 P. La Cour, 'Government Experimental Windmill at Askor', *M.P.I.C.E.*, Vol. 145, 1901, pp. 387–9; Eggleston & Stoddard, 1987, p. 9; Golding, 1976, pp. 15–16; A. Titley, 'Notes on Old Windmills', *T.N.S.*, Vol. 3, 1922–3, p. 50.
7 *See* Baker, 1985, p. 436, note 6; Major, 1977, pp. 26 & 34.
8 Ball, 1908, p. 326.
9 Brown, 1989, p. 108.
10 Batten, 1930, p. 71; Finch, 1976, p. 268; R. Wailes & J. Russell, 'Windmills in Kent', *T.N.S.*, 1953–5, Vol. 29, pp. 225–6.
11 Wolff, 'Letter'.
12 E.L. Burne, 'Wind Power', *The Engineer*, Vol. 107, 19 March 1909, p. 308.
13 Ball, 1908, p. 326.
14 Burne, 'Wind Power', p. 309.
15 Kennedy, 1910, p. 18.
16 *See* Jager, *Catalogue*, *c.* 1895.
17 Ball, 1908, p. 331; Calvert, 1979, p. 7.
18 Eggleston & Stoddard, 1987, p. 26.
19 Calvert, 1979, p. 97.
20 Baker, 1985, p. 46; Williams, 1978, p. 216.
21 Koeppl, 1982, p. 216.
22 Baker, 1985, p. 46.
23 Vowles & Vowles, 1931, pp. 338–9.
24 *See* Lanchester, 1907.
25 Koeppl, 1982, p. 102.
26 Golding, 1976, p. 3.
27 ibid., p. 4.
28 Koeppl, 1982, p. 17.
29 ibid., p. 216.
30 Golding, 1976, pp. 38–9.
31 'Prinsenmolen'-Committee, no date, p. 41.
32 Koeppl, 1982, p. 106.
33 Baker, 1985, p. 46.
34 Koeppl, 1982, pp. 107–12.
35 Putnam wrote a full account of researching, building and running this wind turbine on Grandpa's Knob in his book *Power from the Wind*, published originally in 1948 but reprinted as G.W. Koeppl, *Putnam's Power from the Wind*, van Nostrand Reinhold, New York, U.S.A., 1982.
36 ibid., p. 44.
37 ibid., p. 118.
38 Golding, 1976, p. 212.
39 Eggleston & Stoddard, 1987, p. 317.
40 ibid., p. 86.
41 Nacfaire, 1988, p. 97.
42 Eggleston & Stoddard, 1987, p. 77.

43 ibid., p. 338.
44 *Power from the Wind*, Hydro Electric Leaflet, Edinburgh.
45 Nacfaire, 1988, p. 69.
46 *Power from the Wind*.
47 Leaflets from Vestas, Denmark.
48 D. Quarton, 'Blowin' in the Wind', *Conservation Now*, Vol. 1, No. 2, June/July 1990, p. 39.
49 Koeppl, 1982, p. 215.

Glossary

Annular sail Sail with a single row of shutters in a circle.
Bedstone The lower, fixed, stone of a pair of millstones.
Bolter A machine for separating flour from bran by power-assisted sifting through a bolting cloth.
Brakewheel The primary gearwheel mounted on the windshaft on the rim of which the band brake acts.
Breast beam Main beam at the front of the mill taking the weight of the windshaft.
Bridge tree Hinged beam supporting a stone spindle.
Buck The body of a post mill.
Burr stone Type of stone for making millstones, imported from France.
Canister *see* poll end.
Cap The movable top of a tower mill which turns on a curb on the top of the tower to bring the sails into the wind.
Common sails The traditional sail, covered with cloth.
Cross A casting mounted on the end of a windshaft to carry the sails on its arms.
Cross trees Main horizontal beams in the sub-structure of a post mill.
Crown tree Horizontal beam which bears the weight of the buck of a post mill on the post.
Curb The track on the top of the masonry tower of a mill on which the cap or head turns.
Damsel Device for agitating the shoe to deliver grain into the stones for milling.
Dressing Separating flour from the rest of the meal.
Dressing Recutting the millstones so that they grind more efficiently.
Edge runner *see* runner stone.
Fantail A sail which automatically turns the mill into the wind; sometimes called a fly.
Governor Automatic device to control the distance between the millstones when milling, or the speed of the mill.
Great spur wheel The gearwheel mounted on the upright shaft which drives the stone nuts.
Hollow post mill A mill in which the drive is taken by a shaft through the central post to the machinery below.
Hopper Grain container mounted over the millstones.

Horse Framework on top of the stone casing supporting the hopper and shoe.
Jog-scry An inclined oscillating sieve.
Lantern gear An early form of gearwheel in which a circle of staves held between two wooden flanges served the purpose of cogs. It was usually associated with a driving wheel fitted with peg-like cogs which meshed with the staves of the lantern gear.
Leading boards Longitudinal boards fixed along the leading sides of sails.
Mill mound Generally any raised ground upon which a windmill was constructed.
Neck journal The front bearing surface of a windshaft, turning in a neck bearing. The journal is of iron, the bearing often of stone, wood or metal.
Overdrift Millstones driven from above.
Patent sails Self-regulating sails with shutters.
Poll end or **Canister** Cast iron fitting on the end of the windshaft into which the sail stocks are fixed. The fitting consists of two iron boxes at right angles to each other; formerly the sail stocks were mortised through the end of the timber windshaft.
Post mill A windmill of which the timber-framed body, carrying the machinery and the sails, turns on a vertical post.
Quant The spindle carrying the stone nut which drives an overdrift stone.
Quarter bars The diagonal timbers in the sub-structure of a post mill which brace the post from the cross trees.
Roller mill A form of corn mill developed in the nineteenth century, in which steel rollers replaced millstones to grind the corn.
Roller reefing sails Sails with roller blinds instead of shutters.
Runner stone The upper, driven stone of a pair of millstones; **edge runner** stones are vertical stones which revolve on a circular base and were used for grinding and crushing in many industrial mills.
Rynd Metal cross let into the runner stone to take the drive.
Sack hoist Windpowered mechanism for raising sacks through the mill.
Scoopwheel A waterwheel used in reverse for draining land.
Sheers The two main timbers extending from breast to tail in the cap of a mill. Also the two timbers below a post mill body.
Shutters The hinged parts of a spring or patent sail.
Side girts Timbers running the full length at the sides of a post mill.
Smock mill Tower mill made with a wooden framework.
Smut machine A type of dressing machine to remove smut from grain.
Spring sails Shuttered sails with the shutters in each sail controlled by springs.
Stocks The timbers to which the sails are fixed. A four-sailed mill has two stocks, each of which is passed half-way through the poll end and wedged in place, and each then carries two sails.
Stone nut The final driven pinion in the drive to the stones. In an overdrift mill this gear is mounted on the quant, in an underdrift mill on the stone spindle.
Stone Spindle An iron spindle on the top of which the runner stone is carried.
Tail pole A timber lever for winding the mill body in the case of a post mill, or cap in the case of a tower mill, manually or with the aid of a winch.
Tentering Adjusting the gap between millstones.
Tower mill Masonry towers of which only the cap or head moves to bring the sails into the wind.
Trundle Wheel Wooden gearwheel with pegs instead of cogs.

Underdrift Millstones driven from below.

Upright shaft or **vertical shaft** The main shaft driven by the wallower and carrying the spur wheel for taking off secondary drives.

Vanes A term sometimes used for the shutters in a sail but more generally for a device to wind American windmills.

Wallower The first driven gear which meshes with the brakewheel.

Whip The timber which is bolted to the stock and to which the bars and laths of the sail frame are fixed.

Winding The operation of turning the head or cap of a tower mill or the body of a post mill so that the sails face squarely into the wind.

Windshaft The shaft carrying the sails on its outer end on which the brakewheel is mounted. It is inclined towards the back of the mill in order to distribute some of the load of the sails into the mill structure.

Wire machine A dressing machine used to separate flour from meal in several qualities by brushing the ground meal through wire gauze.

Bibliography

Agricola, G., *De Re Metallica*, 1556. Trans. H.C. Hoover & L.H. Hoover, Dover, New York, 1950.

Anon, *Wilton Windmill*. Wiltshire County Council & Wilton Windmill Soc., 1979.

Apling, H., *Norfolk Corn and Other Industrial Windmills*. Norfolk Windmills Trust, Norwich, 1984.

Apted, M.R., *Sibsey Trader Mill*. Historic Buildings & Monuments Commission for England, H.M.S.O., 1986.

Armstrong, Col. J., *History of the Navigation of King's Lynn*, 1725.

Badeslade, T., *The History of the Ancient and Present State of the Navigation of the Port of King's Lyn, and of Cambridge, and of the Rest of the Trading Towns in those Parts*. London, 1725.

Baker, T.L., *A Field Guide to American Windmills*. University of Oklahoma Press, 1985.

Ball, R.S., *Natural Sources of Power*. Constable, London, 1908.

Bateson, P., *An Answer to Some Objections of Hatton Berner's Esq.*, 1710.

——*Some Papers Relating to the General Draining of Marshlands in the County of Norfolk with Mr Berner's Objections to the Proposals*, 1710.

Bathe, G., *Horizontal Windmills, Draft Mills and Similar Air-flow Engines*. Philadelphia, 1948.

Batten, M.I., *English Windmills*, Vol. I. Society for the Protection of Ancient Buildings, Architectural Press, London, 1930. Vol. II by Smith, D., 1932.

Bawden, T.A., Garrad, L.S., Qualtrough, J.K. & Scatchard, J.W., *Industrial Archaeology of the Isle of Man*. David & Charles, Newton Abbot, 1972.

Bean, D.J., *Herne Mill. A Technical Description*. Hove, 1987.

Beedell, S., *Windmills*. David & Charles, Newton Abbot, 1975.

Beenhakker, A.J., *Windmills and Watermills in Iceland*. The International Molinological Society, Netherlands, 1976.

Belidor, B.F., *Architecture Hydraulique, ou L'Art de Conduire, d'Elever et de Ménager les Eaux pour les Différents Besoins de la Vie*, Tome II, 1782. Originally published C.A. Jombert, Paris, 1737–53.

Bennett J. & Elton, R., *History of Corn Milling*. Simpkin Marshall, London, 1898–1904.

Besson, J., *Théatre des Instruments Mathématiques*, 1578.

Blith, W., *The English Improver*, London, 1649. New edn, *The English Improver Improved*, London, 1652.

Blom, L.H., *The Tjasker Windmill*. The International Molinological Society, Netherlands, 1975.

Bolam, F.M., *Stuff Preparation for Paper and Board Making*. London, 1965.

Boucher, C.T.G., *James Brindley Engineer, 1716–1772*. Goose, Norwich, 1968.

——*John Rennie, 1761–1821, The Life and Work of a Great Engineer*. Manchester University Press, 1963.

British Wind Energy Association, *Wind Energy for the Eighties*. Peter Perigrinus, Stevenage, 1982.

Brown, R.J., *Windmills of England*. Robert Hale, London, 1976. Reprinted, 1989.

Brunnarius, M., *The Windmills of Sussex*. Philimore, London, 1979.

Buchanan, R., *Practical Essays on Millwork and Other Machinery with Notes and Additions by Thomas Tredgold Revised into a Third Edition by G. Rennie*. J. Weale, 3rd, edn, London, 1841.

Buchanan, R.A. & Cossons, N., *The Industrial Archaeology of the Bristol Region*. David & Charles, Newton Abbot, 1969.

Buckland, S., *Lee's Patent Windmill, 1744–1747*. Wind & Watermill Section, Society for Protection of Ancient Buildings, London, 1987.

Burrell, A., *Exceptions Against Sir Cornelius Virmudens Discourse for the Draining of the Great Fennes*. London, 1642.

Butt, J., *The Industrial Archaeology of Scotland*. David & Charles, Newton Abbot, 1967.

Caarten, A. Bicker, *Middeleeuwse Watermolens in Hollands Polderland, 1407/8 – rondom 1500*. Stitching Uitgeverij Noord-Holland, Wormerveer, Netherlands, 1990.

Calvert, N.G., *Windpower Principles, Their Application on the Small Scale*. C. Griffin & Co. Ltd., London, 1979.

Casaubon, I., *Ephermerides*, 1611. New edn., ed. J. Russell, Oxford, 1850.

Caton, P.F.G., *Climatological Memorandum 79: Maps of Hourly Mean Wind Speed Over the United Kingdom, 1965–73*. Meteorological Office, Bracknell, 1976.

Chatterton, A., *Agricultural and Industrial Problems in India*. G.A. Natesan, Madras, India, 1904.

Clapham, Sir J., *A Concise Economic History of Britain from the Earliest Times to 1750*. Cambridge University Press, 1940, reprinted 1951.

Clarke, A., *Windmill Land*. J.M. Dent, London, 1916, reprint G., Kelsall, Littleborough, 1986.

Cochran, T.C., & Miller, W., *The Age of Enterprise. A Social History of Industrial America*, revised edn. Harper & Row, New York, 1961.

Comp, T.A., & Hoeft, K.S., *Long Island Wind and Tide Mills. An Interim Report*. Historic American Engineering Record, Washington D.C., 1976.

Copeland, R., *A Short History of Pottery Raw Materials and the Cheddleton Flint Mill*, Cheddleton Flint Mill Heritage Trust, Leek, 1972.

Corfield, M.C. (Ed.), *A Guide to the Industrial Archaeology of Wiltshire*. Wiltshire County Council Library & Museum Service, Trowbridge, 1978.

Coulthard, A.J., & Watts, M., *Windmills of Somerset and the Men who Worked Them*. Research Publishing Co., London, 1978.

Crocker, A., & Kane, M., *The Diaries of James Simmons, Papermaker of Haslemere, 1831–1868*. Tabard Private Press, Oxshott, 1990.

D'Acres, R., *The Art of Water-Drawing*. H. Brome, London, 1659. Reprinted for the Newcomen Society by Heffers, Cambridge, 1930.

Darby, H.C., *The Draining of the Fens*. Cambridge University Press, 1940.

——*The Medieval Fenland*. Cambridge University Press, 1940.

Daumas, M. (Ed.), *Histoire Générale des Techniques, Tome III, L'Expansion du Machinisme*. Presses Universitaires de France, Paris, 1968.

Defoe, D., *A Tour Through the Whole Island of Great Britain*, 1724.

de Iongh, J., *Van Gelder Zonen, 1784–1934*. de Erven F. Bohn, Haarlem, Netherlands, 1934.

de Little, R.J., *The Windmill Yesterday and Today*. J. Baker, London, 1972.

Dempsey, G.D., *Rudimentary Treatise on the Drainage of Districts and Low Lands*. J. Weale, London, 1854.

den Besten, J., Pouw, G.J., & de Vries, G., *Nieuw Utrechts Molenboek*. Uitgeverij Matrijs, Utrecht, Netherlands, 1991.

Desaguliers, J.T., *A Course of Experimental Philosophy*, 3rd edn. London, 1743.

Dickinson, H.W., & Jenkins, R., *James Watt and the Steam Engine*, 1927. Reprint Moorland Publishing Co. Ltd., Ashbourne, 1981.

Dickinson, H.W., & Titley, A., *Richard Trevithick, The Engineer and the Man*. Cambridge University Press, 1934.

Diderot, D., *Recueil de Planches, sur les sciences, les arts libéraux, et les arts Méchaniques, avec leur Explication*. Briasson, Paris, France, 1763. Copy by Dover Publications, New York, 1959.

Dobber, W., & de Vries, G., *Het Besonder Creckwerck, The Special Crank Work*. Uitgeest, Netherlands, *c.* 1990.

Dodson, W., *The Designe for the Perfect Draining of the Great Level of the Fens*. London, 1665.

Dolman, P., *Lincolnshire Windmills, A Contemporary Survey*. Lincolnshire County Council, Lincoln, 1986.

——*Windmills in Suffolk*. Suffolk Mills Group, 1978.

Douch, H.L., *Cornish Windmills*. O. Blackford, Truro, 1963.

Duckham, B.F., *A History of the Scottish Coal Industry*, Vol. 1, *1700–1815. A Social and Industrial History*. Kelly, New York, 1970.

Eggleston, D.M. & Stoddard, F.S., *Wind Turbine Engineering and Design*. Van Nostrand Reinhold, New York, 1987.

Emerson, W., *The Principles of Mechanics Explaining and Demonstrating the General Laws of Motion*, 1st, edn, 1754, 4th, edn, G.G. & J. Robinson, London, 1794.

Encyclopaedia Britannica, Cambridge University Press, 11th edn, 1910–11, Encyclopaedia Britannica Co., London. 13th edn, 1926.

Fairbairn, W., *Treatise on Mills & Millwork*. Longman, Green, Longman & Roberts, London, 1861.

Farries, K.G. & Mason, R.T., *The Windmills of Surrey and Inner London*. Skilton, London, 1966.

Farey, J., *A Treatise on the Steam Engine*. Longman, Rees, Orme, Brown & Green, London, 1827.

Ferguson, J., *Lectures on Select Subjects in Mechanics, Hydrostatics, Hydraulics, Pneumatics, and Optics*, 1773. New edn, A. Strahan, London, 1803.

Finch, W.C., *Watermills and Windmills. A Historical Survey of their Rise, Decline and Fall as Protrayed by those of Kent*. C.W. Daniel, 1933. New edn, A. Cassell, Sheerness, 1976.

Flint, B., *Windmills of East Anglia*. F.W. Pawsey, Ipswich, 1977.
——*Suffolk Windmills*. Boydell Press, Woodbridge, 1979.
Focsa, G., Ed., *Muzeul Satului, Studii si Cercetari*. Bucharest, Romania, 1971.
Forbes, R.J., *Man the Maker, A History of Technology and Engineering*. Constable & Co. Ltd., London, 1958.
——*Studies in Ancient Technology*. E.J. Brill, Leyden, Netherlands, 1955.
Freese, S., *Windmills and Millwrighting*. Cambridge University Press, 1957. Reprint, David & Charles, Newton Abbot, 1971.
Fussell, G.E., *The History of the Farmer's Tools*, 1952. Reprint, Bloomsbury Books, London, 1985.
Galloway, R.L., *A History of Coal Mining in Great Britain*. Macmillan, London, 1882. Reprint, David & Charles, Newton Abbot, 1969.
Gaucheron, A. & Major, J.K., *The Eolienne Bollée*. The International Molinological Society, Reading, 1985.
Gimpel, J., *The Medieval Machine*. Futura Publications Ltd., London, 1979.
Golborne, J., *Report of James Golborne of the City of Ely, Engineer*, 1792.
Golding, E.W., *The Generation of Electricity by Wind Power*. E. & F. Spon, London, 1955. Reprint, 1976.
Gooch, W., *The General View of the County of Cambridge*. Board of Agriculture, London, 1813.
Gray, A., *The Experienced Millwright; or a Treatise on the Construction of Some of the Most Useful Machines with the Latest Improvements*, A. Constable, Edinburgh, 2nd edn, 1806.
Gregory, G., *Dictionary of Arts and Sciences*. London, 1807.
Gregory, R., *East Yorkshire Windmills*. Skilton, Cheddar, 1985.
Griffin, A.R., *Coalmining*. Longman, London, 1971.
——*The British Coalmining Industry, Retrospect and Prospect*. Moorland Publishing Co., Hartington, 1977.
Griffiths, R.T., *Industrial Retardation in the Netherlands, 1830–1850*. Martinus Nijhof B.V., The Hague, Netherlands, 1979.
Guise, B. & Lees, G., *Windmills of Anglesey*. Attic Books, Builth Wells, 1992.
Harte, J.H., *Volledig Molenboek*. A. van der Mast, Gorinchem, Holland, 1849. Reprint, Buitenpost, Netherlands, 1979.
Hartlib, S., *A Discourse of Husbandrie Used in Brabant and Flanders*. W. Du-Gard, London, 1650.
Harverson, M., *Persian Windmills*. The International Molinological Society, Reading, 1991.
Hasted, E., *The History and Topographical Survey of the County of Kent*. Simmons & Kirkby, Canterbury, 1790.
Heathcote, J., *The Scoopwheel and Centrifugal Pump*. London, 1887.
Heathcote, J.M., *Reminiscences of Fen and Mere*. London, 1876.
Helmers, R., *Van Haver Tot Gort, 150 Tekeningen van één Molen*. Uitgave Profiel, Bedum, Netherlands, *c.*, 1990.
Henning, P., *Windmills in Sussex, A description of the Construction and Operation of Windmills, exemplified by Up-to-Date Notes on the Still Existing Windmills in Sussex with Photographic Illustrations*. Daniel, London, 1936.
Hills, R.L., *Machines, Mills and Uncountable Costly Necessities. A Short History of the Drainage of the Fens*. Goose, Norwich, 1967.
——*Papermaking in Britain, 1488–1988, A Short History*. Athlone Press, London, 1988.

——*Power in the Industrial Revolution*. Manchester University Press, 1970.
——*Power from Steam, A History of the Stationary Steam Engine*. Cambridge University Press, 1989.
Holt, R., *The Mills of Medieval England*. B. Blackwell, Oxford, 1988.
Hoogerworst, N. & van 't Land, G., *Why Windmills Applied, Socio-economic Determinants of Windmill Use for Irrigation on the Lasithi Plateau in Crete*. Wageningen, Netherlands, 1983.
Hopkins, R.T., *Old Watermills and Windmills*. P. Allen, London, 1930. New edn, E.P. Publishing, Wakefield, 1976.
Hounshell, D.A., *From the American System to Mass Production, 1800–1932. The Development of Manufacturing Technology in the United States*. Johns Hopkins University Press, Baltimore, 1984.
Household, H., *The Thames & Severn Canal*. David & Charles, Newton Abbot, 1969.
Hughes, W.J., *A Century of Traction Engines*. Percival Marshall, London, 1959.
Hunter, D., *Papermaking, The History and Technique of an Ancient Craft*. Dover Publications, New York, 1978 edn.
Hunter, L.C., *A History of Industrial Power in the United States, 1780–1930, Vol. 1, Waterpower in the Century of the Steam Engine*. University Press of Virginia, Charlottesville, 1979.
Ireland, S., *Picturesque Views on the River Medway with Observations on the Works of Art in its Vicinity*. T. & J. Egerton, London, 1793.
Jager, C.J., *Illustrated Catalogue of Windmills, Tanks and Pumps as Applied to Water Supply Systems, also Windmills Adapted for Power*. C.J. Jager & Co., Boston, *c*, 1895.
Jevons, W.S., *The Coal Question: An Inquiry Concerning the Progress of the Nation, and the Probable Exhaustion of our Coal-mines*. Macmillan, London, 1865.
Kealey, E.J., *Harvesting the Air, Windmill Pioneers in Twelfth Century England*. Boydell Press, Woodbridge, 1987.
Keller, A.G., *A Theatre of Machines*. Chapman & Hall, London, 1964.
Kemp, P., *The History of Ships*. Orbis Publishing, London, 1978. Reprint, 1983.
Kempe, H.R. (Ed.), *The Engineer's Year Book*. Crosby Lockwood & Son, 1929.
Kennedy, R., *The Book of Modern Engines: A Practical Work on Prime Movers and the Transmission of Power: Steam, Electric, Water, Gas and Hot Air*. Caxton Pub. Co. Ltd., London, 1910.
Kinderley, N., *The Ancient and Present State of the Navigation of the Towns of Lyn, Wisbech, Spalding and Boston*. London, 1751.
Kirkus, M., *The Records of the Commissioners of Sewers in the Parts of Holland, 1547–1603*. Ruddock & Sons, Lincoln, 1959.
Klemm, F., *A History of Western Technology*. Allen & Unwin, London, 1959.
Koeppl, G.W., *Putnam's Power from the Wind*. van Nostrand Reinhold, New York, 1982.
Kölker, A.J., *Molens in de Banne Uitgeest*. Uitgave Stichting Uitgeester en Akersloter Molens, Netherlands, 1990.
Labelye, C., *The Result of a View of the Great Level of the Fens taken in July 1745*. London, 1745.
Lambert, J.M., *The Making of the Broads*. Royal Geographical Society Series, London, 1960.
Lancashire, S., Kenna J. & Fraenkel, P., *Windpumping Handbook*. IT Publications, London, 1987.
Lanchester, F.W., *Aerodynamics*. London, 1907.

Landels, J.G., *Enginering in the Ancient World*. University of California Press, Berkeley, 1978.

Lardner, D., *Handbook of Natural Philosophy, Hydrostatics and Pneumatics*, new edn. rewritten by B. Loewy. Lockwood, London, 1874.

le Gouriérès, D., *Wind Power Plants, Theory and Design*. Pergamon Press, Oxford, 1982.

Lewis, M.J.T., *Early Wooden Railways*. Routledge & Kegan Paul, London, 1970.

Linperch, P., *Architectura Mechanica, of Moolen-Boek*. J. Cövens, Amsterdam, 1727. Reprint, P.N. van Kampen & Zoon, Amsterdam, Netherlands, 1969.

McDermott, R. & R., *The Standing Windmills of East Sussex*. Betford, Worthing, 1978.

——*The Standing Windmills of West Sussex*. Betford, Worthing, 1978.

McGuigan, D., *Small Scale Wind Power*. Prism Press, Dorchester, 1978.

MacLeod, C., *Inventing the Industrial Revolution, The English Patent System, 1660–1800*. Cambridge University Press, 1988.

McNeil, I. (Ed.), *An Encyclopeadia of the History of Technology*. Routledge, London, 1990.

——*Joseph Bramah, A Century of Invention, 1749–1851*. David & Charles, Newton Abbot, 1968.

Mais, S.P.B., *England of the Windmills*. J.M. Dent, London, 1931. Reprint, E.P. Publishing, Wakefield, 1978.

Major, J.K., *The Windmills of John Wallis Titt*. International Molinological Society, Netherlands, 1977.

Mann, R.D., *How to Build A 'Cretan Sail' Windpump for use in low-speed wind conditions*. IT Publications, London, 1979.

Mathé, J., *Leonardo's Inventions*. Miller Graphics, Geneva, Switzerland, 1980.

Maxwell, G., *The General View of the County of Huntingdon*. Board of Agriculture, London, 1793.

Mayr, O., *The Origins of Feedback Control*. M.I.T. Press, Cambridge, MA, 1970.

Meteorological Office, *A Course in Elementary Meteorology*. H.M.S.O., London, 1978. 2nd imp., 1981.

Miller, S.H., *A Handbook to the Fenland*. Simpkin & Marshall, London, 1889.

Miller, S.H., & Skertchly, S.B.J., *The Fenland Past and Present*. Leach & Son, Wisbech, 1878.

Morgan Rees, D., *Industrial Archaeology of Wales*. David & Charles, Newton Abbot, 1975.

Mott, R.A., *Henry Cort, The Great Finer, Creator of Puddled Iron*. The Metals Society, London, 1983.

Moxon, J.H.H., *Fenland Floods and the Lower Ouse*. Cambridge, 1878.

Nacfaire, H. (Ed.), *Grid-Connected Wind Turbines*. Commission of the European Communities, Brussels, Elsevier Applied Science, London, 1988.

Neale, T., *The Ruinous State of the Parish of Manea*, 1748.

Needham, J., *Science and Civilization in China*, Vol. I, *Introductory Orientations*. Cambridge University Press, 1954. Vol. IV, *Physics and Physical Technology, Part II, Mechanical Engineering*. Cambridge University Press, 1965.

Nickalls, J., *Report upon the Consequences which the New Cut from Eau-Brink would be Attended with . . .*, 1793.

Nijhof, P., *Wind Molens in Nederland*. Uitgeverij Waander, Zwolle, Netherlands, *c.* 1980.

Notebaart, J.C., *Windmühlen der Stand der Forschung über das Vorkommen und der Ursprung*. Mouton Verlag, The Hague, Netherlands, 1972.
Pels, C. & Voorn, H., *World's Last Wind Paper-mill at Westzaan, The Schoolmaster*, Van Gelder Papier, Amsterdam, Netherlands, no date.
Pouw, G.J., *Wieksystemen voor Polder- en Industriemolens*. Kluwer Technische Boeken B.V., Deventer, Netherlands, 1981.
Priestley, J., *The Historical Account of the Navigable Rivers, Canals and Railways throughout Great Britain*, 1831. Reprint, David & Charles, Newton Abbot, 1969.
'Prinsenmolen'-Committee, *Research Inspired by the Dutch Windmills*. Veenman en Zonen, Wageningen, Netherlands, no date.
Pryce, W., *Mineralogia Cornubiensis; A Treatise on Minerals, Mines and Mining*. J. Phillips, London, 1778.
Putnam, P.C., *Power from the Wind*, *see* Koeppl.
Ramelli, A., *The Various and Ingenious Machines of Agostino Ramelli*. Paris, France, 1588. Dover Pub., New York, 1976.
Rankine, W.J.M., *A Manual of the Steam Engine and Other Prime Movers*. Griffin, Bohn, London, 2 edn, 1861.
Rees, A. (Ed.), *The Cyclopaedia; or Universal Dictionary of Arts, Sciences and Literature*. London, 1819. Reprint David & Charles, Newton Abbot, 1972.
Rennie, G., *Illustrations of Mill Work and Other Machinery together with Tools of Modern Invention; Atlas to the new Edition of Buchanan's Work*. J. Weale, London, 1841.
Reynolds, J., *Windmills and Watermills*. H. Evelyn, London, 1970. Reprint, 1974.
Reynolds, T.S., *Stronger than a Hundred Men, A History of the Vertical Waterwheel*. Johns Hopkins University Press, Baltimore, 1983.
Rhodes, P.S., *Guide to the Ballycopeland Windmill*. H.M.S.O., Belfast, 1962.
Rolt, L.T.C. & Allen, J.S., *The Steam Engine of Thomas Newcomen*. Moorland Publishing Co., Hartington, 1977.
Schofield, R.E., *The Lunar Society of Birmingham, A Social History of Provincial Science and Industry in Eighteenth Century England*. Clarendon Press, Oxford, 1963.
Scott, M., *The Restoration of Windmills and Windpumps in Norfolk*. Norfolk Windmills Trust, Norwich, 1977.
Shaw, J., *Water Power in Scotland, 1550–1870*. J. Donald Publishers, Edinburgh, 1984.
Shorter, A.H., *Paper Mills and Paper Makers in England, 1495–1800*. Paper Publications Society, Hilversum, Netherlands, 1957.
Simon, A., *The Simon Engineering Group*. Simon Engineering Group, Stockport, 1953.
Singer, C. (Ed.), *A History of Technology*, Vols 2 and 3. Clarendon Press, Oxford, 1956–7.
Sipman, A., *Molenbouw Het Staande Werk van de Boven Kruiers*. de Walburg Pers, Zutphen, Netherlands, 1975.
Skempton, A.W. (Ed.), *John Smeaton, F.R.S.* T. Telford Ltd., London, 1981.
Skilton, C.P., *British Windmills and Watermills*. Collins, London, 1948.
Smeaton, J., *Experimental Enquiry Concerning the Natural Powers of Wind and Water to Turn Mills and Other Machines Depending on a Circular Motion*, I. & J. Taylor, London, 2nd edn, 1796.
Smiles, S., *Industrial Biography, Ironworkers and Tool Makers*. J. Murray, London, 1863. Reprint, David & Charles, Newton Abbot, 1967.
——(Ed.), *James Nasmyth, Engineer, An Autobiography*. J. Murray, London, 1897.

——*Lives of the Engineers: Early Engineering: Vermuyden, Myddleton, Perry, James Brindley*. J. Murray, London, 1874.

——*Lives of the Engineers; Harbours, Lighthouses, Bridges: Smeaton and Rennie*. J. Murray, London, 1874.

Smith, A.C., *Drainage Windmills of the Norfolk Marshes*, Smith, Stevenage, new edn, 1990.

Smith, D.M., *The Industrial Archaeology of the East Midlands*. David & Charles, Dawlish, 1965.

Snelling, J.M., *St. Benet's Abbey, Norfolk*, revised W.F. Edwards. Diocesan Offices, Norwich, 1983.

Stevin, 'Introduction to Stevin's Essay on the Drainage-mills', typescript.

Stokhuyzen, F., *The Dutch Windmill*. C.A.J. van Dishoeck, Bussum, Netherlands, 1962.

Stone, T., *The General View of the County of Lincoln*. Board of Agriculture, London, 1794.

Stott, P.H. (Ed.), *Long Island, An Inventory of Historic Engineering and Industrial Sites*. Historic American Engineering Record, Washington DC, *c*. 1975.

Stukeley, W., *Memoirs of Sir Isaac Newton's Life*, 1752. Reprint, Taylor & Francis, London, 1936.

Summers, D., *The Great Level, A History of Drainage and Land Reclamation in the Fens*. David & Charles, Newton Abbot, 1976.

Templeton, W., *The Millwright and Engineer's Pocket Companion*. Simpkin Marshall, London, 1856.

Thomson, A.G., *The Paper Industry in Scotland*. Scottish Academic Press, Edinburgh, 1974.

Torrey, V., *Wind-Catchers, American Windmills of Yesterday and Tomorrow*. Stephen Greene Press, Brattleboro, Vermont, 1976.

Trevithick, F., *Life of Richard Trevithick*. E. & F.N. Spon, London, 1872.

Valls, O., *The History of Paper in Spain, X–XIV Centuries*. Empresa Nacional de Celulosas, Madrid, 1978.

van Dijk, H.J. & Goedhart, P.D., *Windpumps for Irrigation*. C.W.D., Amsterdam, 1990.

van der Pols, K., *De Ontwikkeling van het Vatweropuoewerktig in Nederland*. Delftse Universitaire pers, Netherlands, no date.

van der Woude, A.M., *Het Noorderkwarteer*. HES Uitgeveers, Utrecht, Netherlands, 1983.

Van Heiningen, H., *Diepers, en Delvers, Geschidenis van de Zand-en Grindbaggeraas*, Walburg Pers, Netherlands, 1991.

van Natrus, L., Polly, J. & Van Vuuren, C., *Groot Volkomen Moolenboekk*. J. Cövens, Amsterdam, 1734; and further volume, *Groot Algemeeen Molenboek, alsmede de Sluys*, 1736. Reprint P.N. Van Kamper & Zoon, Amsterdam, Netherlands, 1969.

van Veen, J., *Dredge, Drain, Reclaim, The Art of a Nation*. M. Nijhof, The Hague, Netherlands, 1962.

Varchmin, J. & Radkau, J., *Kraft, Energie und Arbeit; Energie und Gesellschaft*. Deutsches Museum, Munich, 1988.

Verantius, F., *Machinae Novae*. Venice, Italy, *c*. 1620.

Voller, W.R., *Modern Flour Milling*. Gloucester, 3rd edn, 1897.

Vowles, H.P. & Vowles M.W., *The Quest for Power from Prehistoric Times to the Present Day*. Chapman & Hall, London, 1931.

Wailes, R., *Berney Arms Windmill*. H.M.S.O., London, 1957. 2nd edn, English Heritage, London, 1982; 3rd edn, 1990.
——*The English Windmill*. Routledge & Kegan Paul, London, 1954.
——*Saxtead Green Post Mill, Suffolk*. English Heritage, London, 1984.
Walker, N. & Craddock, T., *The History of Wisbech and the Fens*. R. Walker, Wisbech, 1849.
Wallis Titt, J., *Catalogue of Wind Engines and Pumps*. Warminster, *c*. 1905.
Warburton, E., *Life of Horace Walpole*. London, 1851.
Watts, M., *Corn Milling*. Shire Publications, Aylesbury, 1983.
——*Wiltshire Windmills*. Wiltshire Library & Museum Service, Trowbridge, 1980.
Weatherill, L., *The Pottery Trade and North Staffordshire*. 1660–1760. Manchester University Press, 1971.
Webb, W.P., *The Great Plains*. Grosset & Dunlap, New York, 1931. Reprinted 1976.
Wells, S., *The History of the Drainage of the Great Level of the Fens, Called Bedford Level*. R. Pheney, London, 1830.
Wheeler, W.H., *History of the Fens of South Lincolnshire*, 2nd edn. Newcombe, Boston, 1897.
White, L., *Medieval Technology and Social Change*. Clarendon Press, Oxford, 1962.
Williams, M., *The Draining of the Somerset Levels*. Cambridge University Press, 1970.
Williams, T.I. (Ed.), *A History of Technology*. Vol. VI *The 20th Century, c. 1900 to 1950*. Clarendon Press, Oxford, 1978.
Wilson, F.M., *Strange Island: Britain through Foreign Eyes, 1395–1940*. London, 1955.
Wolf, A., *A History of Science, Technology and Philosophy in the Eighteenth Century*, 2nd edn, ed. D. McKie. Allen & Unwin, London, 1962.
Wolff, A.B., *The Windmill as a Prime Mover*, 2nd edn. Wiley, New York, 1885.
Young, A., *The General View of the County of Lincoln*. Board of Agriculture, London, 1799.
Zonca, V., *Novo Teatro di Machine et Edificii*. Padua, Italy, 1607.

Journals, Periodicals, etc.

Annals of Agriculture (*A.A.*)
Cambridge Industrial Archaeology Bulletin (*C.I.A.B.*)
Conservation Now (*C.N.*)
Daedalus, Yearbook of Tekniska Museet, Stockholm, Sweden
The Engineer
Industrial Archaeology Review (*I.A.R.*)
Industriëlle Archaeology, Netherlands (*I.A.N.*)
Journal of the House of Commons (*J.H.C.*)
Journal of Industrial Archaeology (*I.A.*)
Journal of the Royal Agricultural Society (*J.R.A.S.*)
Met Stoom, Netherlands
Minutes of the Proceedings of the Institution of Civil Engineers (*M.P.I.C.E.*)
The Quarterly, The Review of the British Association of Paper Historians
Transactions of the International Molinological Society (*T.I.M.S.*)
Transactions of the Newcomen Society (*T.N.S.*)
Transactions of the Society of Arts (*T.S.A.*)
Windmillers' Gazette, U.S.A. (*W.G.*)

Index

FRANKLIN PIERCE COLLEGE LIBRARY
00080668